Lecture Notes in Mathematics 1200

Springer
Berlin
Heidelberg
New York
Barcelona
Hong Kong
London
Milan
Paris
Singapore
Tokyo

Vitali D. Milman   Gideon Schechtman

Asymptotic Theory of Finite Dimensional Normed Spaces

With an Appendix by M. Gromov
"Isoperimetric Inequalities in Riemannian Manifolds"

Springer

Authors

Vitali D. Milman
Department of Mathematics
Tel Aviv University
Ramat Aviv, Israel

Gideon Schechtman
Department of Theoretical Mathematics
The Weizmann Institute of Science
Rehovot, Israel

Cataloging-in-Publication Data applied for

Die Deutsche Bibliothek - CIP-Einheitsaufnahme

Milman, Vitali D.:
Asymptotic theory of finite dimensional normed spaces / Vitali D. Milman ; Gideon Schechtman. With an appendix Isoperimetric inequalities in Riemannian manifolds / by M. Gromov. - Corr. 2. printing. - Berlin ; Heidelberg ; New York ; Barcelona ; Hong Kong ; London ; Milan ; Paris ; Singapore ; Tokyo : Springer, 2001
(Lecture notes in mathematics ; 1200)
ISBN 3-540-16769-2

Corrected Second Printing 2001

Mathematics Subject Classification (1980): 46B20, 52A20, 60F10

ISSN 0075-8434
ISBN 3-540-16769-2 Springer-Verlag Berlin Heidelberg New York

Springer-Verlag Berlin Heidelberg New York
a member of BertelsmannSpringer Science+Business Media GmbH

http://www.springer.de

Printed in Germany

Typesetting: Camera-ready TEX output by the authors
SPIN: 10797471 41/3142-543210 - Printed on acid-free paper

INTRODUCTION

This book deals with the geometrical structure of finite dimensional normed spaces, as the dimension grows to infinity. This is a part of what came to be known as the Local Theory of Banach Spaces (this name was derived from the fact that in its first stages, this theory dealt mainly with relating the structure of infinite dimensional Banach spaces to the structure of their lattice of finite dimensional subspaces).

Our purpose in this book is to introduce the reader to some of the results, problems, and mainly methods developed in the Local Theory, in the last few years. This by no means is a complete survey of this wide area. Some of the main topics we do not discuss here are mentioned in the Notes and Remarks section. Several books appeared recently or are going to appear shortly, which cover much of the material not covered in this book. Among these are Pisier's [Pis6] where factorization theorems related to Grothendieck's theorem are extensively discussed, and Tomczak-Jaegermann's [T-J1] where operator ideals and distances between finite dimensional normed spaces are studied in detail. Another related book is Pietch's [Pie].

The first major result of the Local Theory is Dvoretzky's Theorem [Dv] of 1960. Dvoretzky proved that every real normed space of finite dimension, say n, contains a $(1+\varepsilon)$-isomorphic copy of the k-dimensional euclidean space ℓ_2^k, for $k = k(\varepsilon, n)$ which increases to ∞ with n (see Chapter 5 for the precise statement). Dvoretzky's original proof was very complicated and understood only by a few people. In 1970 Milman [M1] gave a different proof which exploited a certain property of the Haar measure on high dimensional homogeneous spaces, a property which is now called the concentration phenomenon: Let (X, ρ, μ) be a compact metric space (X, ρ) with a Borel probability measure μ. The concentration function $\alpha(X, \varepsilon)$, $\varepsilon > 0$, is defined by

$$\alpha(X,\varepsilon) = 1 - inf\{\mu(A_\varepsilon);\ \mu(A) \geq \frac{1}{2},\ A \subseteq X\ Borel\}$$

where

$$A_\varepsilon = \{x \in X;\ \rho(x, A) \leq \varepsilon\}\ .$$

It turns out that for some very natural families of spaces, $\alpha(X, \varepsilon)$ is extremely small. For example, it follows from Levy's isoperimetric inequality that for the Euclidean n-sphere S^n, with the geodesic distance ρ and the normalized rotational invariant measure μ,

$$\alpha(S^n, \varepsilon) \leq \sqrt{\frac{\pi}{8}} exp(-\varepsilon^2 n/2)\ .$$

It follows from this inequality (see Chapter 2) that any nice real function on S^n must be very close to being a constant on all but a very small set (the exceptional set being of measure of order smaller than $exp(-\varepsilon^2 n/2)$). This last property is what is called the concentration phenomenon. It has proved to be extremely useful in the study of finite dimensional normed spaces.

Going back to the concentration function, we define a family (X_n, ρ_n, μ_n) of metric probability spaces to be a Levy family if $\alpha(X_n, \varepsilon_n \; diam \; X_n) \underset{n\to\infty}{\longrightarrow} 0$ ($diam \; X_n$ is the diameter of X_n). Chapter 6 below contains a lot of examples of such natural families. Many of these examples have deep applications in the Local Theory. It is usually quite a difficult task to establish that a certain family is a Levy family, the methods are different from one example to the other and come from diverse areas (including methods from differential geometry, estimation of eigenvalues of the Laplacian, large deviation inequalities for martingales, isoperimetric inequalities). Levy families, the concentration phenomenon and their applications to the asymptotic theory of normed spaces are the main topics of the first part of this book. We have already mentioned one application, namely Dvoretzky's Theorem. In the same direction we deal with estimation of the dimension of euclidean subspaces of various large families of normed spaces, that is, with the evaluation of the function $k(\varepsilon, n)$ mentioned above when restricted to some (wide) families of normed spaces. (This study originated in [M1] and [F.L.M.].) Here is an example: There exists a function $c(\varepsilon) > 0$, $\varepsilon > 0$, such that if $X_n = (\mathbb{R}^n, \|\cdot\|)$ is a family of normed spaces, then either for some $\alpha > 0$ and any $\varepsilon > 0$ and n, X_n contains a $(1+\varepsilon)$-isomorphic copy of ℓ_2^k with $k = [c(\varepsilon)n^\alpha]$, or for any integer k and any $\varepsilon > 0$, there is an n such that X_n contains a $(1+\varepsilon)$-isomorphic copy of ℓ_∞^k. (The proof of this result uses, besides the concentration phenomenon, also the notions of type and cotype introduced below.)

We also deal, in the first part, with packing high dimensional ℓ_p^k, $1 < p < 2$, spaces into ℓ_1^n (Chapter 7) as well as packing spaces with special structure (unconditional or symmetric bases) into general normed spaces (Chapter 10).

The second part of the book revolves around the notions of type and cotype and the relation of these notions to the geometry of normed spaces.

For $1 \le p \le 2 \le q \le \infty$ the type p constant (resp. cotype q constant) of X denoted $T_p(X)$ $(C_q(X))$ is the smallest constant in the inequality

$$\left(\underset{\varepsilon_i=\pm 1}{\mathrm{Ave}} \left\|\sum_{i=1}^k \varepsilon_i x_i\right\|^2\right)^{1/2} \le T \left(\sum_{i=1}^k \|x_i\|^p\right)^{1/p}$$

$$\left(\left(\sum_{i=1}^k \|x_i\|^q\right)^{1/q} \le C \left(Ave\left\|\sum_{i=1}^k \varepsilon_i x_i\right\|^2\right)^{1/2} \right)$$

for all k and $x_1, \ldots, x_k \in X$.

These notions were introduced by Hoffmann-Jorgensen [H-J] for the study of limit theorems for vector valued random variables, and were studied extensively by Maurey and Pisier ([M.P.] in particular) in connection with the geometry of normed spaces. They have proved to be a very important tool in the Local Theory. In particular, Krivine, Maurey and Pisier showed that for $p_X = sup\{p;\ T_p(X) < \infty\}$ (resp. $q_X = inf\{q;\ C_q(X) < \infty\}$), $\ell^n_{p_X}$ (resp. $\ell^n_{q_X}$) are $(1+\varepsilon)$-isomorphic to subspaces of X for all n and $\varepsilon > 0$ (here X is infinite dimensional; there is also a corresponding statement for finite dimensional spaces). We present a proof (somewhat different from previous proofs) of this theorem in Chapters 12 and 13. (Chapter 11 deals with some infinite dimensional combinatorial methods needed in the sequel.)

Chapter 14 is devoted to the work of Pisier [Pis1] estimating the norm of one specific projection called the Rademacher projection. This is closely related to the relation between the type p constant of X and the cotype q constant of X^* $(\frac{1}{p}+\frac{1}{q}=1)$. It also has applications to finding well complemented euclidean sections in normed spaces. These applications due to Figiel and Tomczak-Jaegermann [F.T.] are discussed in Chapter 15.

The book also contains five appendices, the first of which is written by M. Gromov and gives an introduction to the theory of isoperimetric inequalities on riemannian manifolds. It is written in a way understandable to the non-expert (in Differential Geometry). This appendix contains also results which were not published elsewhere (in particular – the Gromov-Levy isoperimetric inequality). We are indebted to M. Gromov for this excellent addition to our book.

CONTENTS

Appendices

PART I: THE CONCENTRATION OF MEASURE PHENOMENON IN THE THEORY OF NORMED SPACES

1. PRELIMINARIES.

In this chapter we present some preliminary material on invariant measures needed in later chapters. Not all the details of the proofs are given.

1.1. Let (M, ρ) be a compact metric space and let G be a group whose members act as isometries on M, i.e. for $g \in G$, $t, s \in M$, $\rho(gt, gs) = \rho(t, s)$.

We begin by introducing the Haar measure. The proof we give is due apparently to W. Maak (see [Do]).

THEOREM: *There exists a regular measure μ on the Borel subsets of M which is invariant under the action of members of G, i.e., $\mu(A) = \mu(gA)$ for all $A \subseteq M$, $g \in G$. Alternatively, $\int f(t)d\mu(t) = \int f(gt)d\mu(t)$ for all $g \in G$ and $f \in C(M)$. ($C(M)$ is the linear space of all real continuous functions on M).*

PROOF: For each $\varepsilon > 0$ let N_ε be a minimal ε-net in M, i.e., $\cup_{t \in N_\varepsilon} B(t, \varepsilon) = M$ and $n_\varepsilon = |N_\varepsilon|$, the cardinality of N_ε, is minimal among all sets with this property. ($B(t, \varepsilon) = \{s \in M; \rho(t, s) \leq \varepsilon\}$).

For $f \in C(M)$ define

$$\mu_\varepsilon(f) = n_\varepsilon^{-1} \sum_{t \in N_\varepsilon} f(t) .$$

Since $\{\mu_\varepsilon\}_{\varepsilon > 0}$ is a uniformly bounded set of linear functionals on $C(M)$, it follows (see e.g. [D.S.]) that, for some sequence $\varepsilon_i \underset{i \to \infty}{\longrightarrow} 0$,

$$\mu_{\varepsilon_i}(f) \underset{i \to \infty}{\longrightarrow} \mu(f)$$

for all $f \in C(M)$ where $\mu(\cdot)$ is a linear positive functional on $C(M)$ with $\mu(1) = 1$. That is, μ is given by a regular Borel probability measure (see [D.S.] again).

Next we want to show that the measure μ is uniquely determined, i.e., if μ'_ε is defined using a different minimal ε-net, then for the same sequence ε_i, $\mu'_{\varepsilon_i}(f) \to \mu(f)$. If N'_ε is another minimal ε-net in M then we claim: there exists a one to one and onto map $\varphi: N_\varepsilon \to N'_\varepsilon$ with $\rho(t, \varphi(t)) \leq 2\varepsilon$ for all $t \in N_\varepsilon$.

To show this we use a combinatorial result known as the "marriage Theorem" (see e.g.[Do]). Let us say that $t \in N_\varepsilon$ and $s \in N'_\varepsilon$ are acquainted if $B(t,\varepsilon) \cap B(s,\varepsilon) \neq \emptyset$. Then the members of any subset K of N_ε are collectively acquainted with a subset L of N'_ε of at least as many elements as K. Indeed, let $L = \{s \in N'_\varepsilon; B(s,\varepsilon) \cap (\cup_{t\in K} B(t,\varepsilon)) \neq \emptyset\}$ then $|L| \geq |K|$ since otherwise $L \cup (N_\varepsilon \backslash K)$ is an ε-net with less then $|N_\varepsilon|$ elements. The marriage Theorem states that in such a situation (where every subset of N_ε is jointly acquainted with at least as many elements of N'_ε) there is a one to one map $\varphi: N_\varepsilon \to N'_\varepsilon$ with t and $\varphi(t)$ acquainted for all $t \in N_\varepsilon$. This translates back into $B(t,\varepsilon) \cap B(\varphi(t),\varepsilon) \neq \emptyset$, i.e., $\rho(t,\varphi(t)) \leq 2\varepsilon$. It follows that, if μ'_ε is defined using N'_ε in an analogous way to μ_ε, then

$$|\mu_\varepsilon(f) - \mu'_\varepsilon(f)| \leq \frac{1}{n_\varepsilon} \sum_{t\in N_\varepsilon} |f(t) - f(\varphi(t))| \leq \omega(2\varepsilon)$$

where $\omega(\varepsilon) = sup\{|f(t) - f(s)|; \rho(t,s) \leq \varepsilon\}$ is the modulus of continuity of f. Thus, $\lim_{i\to\infty} \mu'_{\varepsilon_i}(f)$ exists and is equal to $\mu(f)$.

It remains to show that μ is invariant under G. Let $g \in G$ and let $N'_\varepsilon = (gt)_{t\in N_\varepsilon}$. Then N'_ε is a minimal ε-net and

$$\begin{aligned}\mu(f \circ g) &= \lim_{i\to\infty} \frac{1}{n_{\varepsilon_i}} \sum_{t\in N_{\varepsilon_i}} f(gt) = \lim_{i\to\infty} \frac{1}{n_{\varepsilon_i}} \sum_{s\in N'_{\varepsilon_i}} f(s) \\ &= \lim_{i\to\infty} \mu'_{\varepsilon_i}(f) = \mu(f) .\end{aligned}$$

□

1.2. If the action of G on M is transitive, i.e. for all $t, s \in M$ there exists a $g \in G$ such that $gt = s$, then M is called an *homogeneous space* of G.

Fix a $t \in M$ and let

$$G_o = \{g \in G; gt = t\}$$

then G_o is a subgroup of G (called an *isotropic subgroup*) and $M = G/G_o$ where $s \in M$ is identified with the equivalence class gG_o for g such that $gt = s$.

To illustrate the definition we give a simple example : Fix an inner product $(\cdot,\cdot)$ on $\mathbb{R}^n$, let $|x| = (x,x)^{1/2}$ for $x \in \mathbb{R}^n$ and let $G = O_n$ be the orthogonal group on $(\mathbb{R}^n, \{\cdot,\cdot\})$. We identify O_n with the set of n tuples $(e_1,\ldots e_n)$ of orthonormal vectors (Fix one orthonormal basis $(e_1^o,\ldots,e_n^o)$ then any orthogonal operator A uniquely determines another such n-tuple: $(Ae_1^o,\ldots,Ae_n^o)$). Let $M = S^{n-1} = \{x \in \mathbb{R}^n; |x| = 1\}$ and $\varphi: O_n \to S^{n-1}$ be defined by $\varphi(e_1,\ldots,e_n) = e_1$. Then clearly for any $t \in S^{n-1}, \varphi^{-1}(t)$ can be identified with O_{n-1}. So $S^{n-1} = O_n/O_{n-1}$.

1.3. THEOREM: *If (M,ρ) is a compact metric homogeneous space of the group G then the measure of Theorem 1.1. is unique up to a constant.*

PROOF: Define a semi metric on G by $d(g,h) = sup_{t\in M}\rho(gt,ht)$. Identifying elements whose distance apart is zero, we get a group H which still acts as isometries on M and also on itself. (there are two ways in which H acts on itself - we choose multiplication on the right $hh' = h' \cdot h$ where " $\cdot$ " is the multiplication in the group). One checks that H is compact (actually in all our applications $G = H$ will be given as a compact group). Let μ on M and ν on H be measures invariant under the action of G. Then, for all $f \in C(M)$,

$$\nu(1)\mu(f) = \int_G \int_M f(gt)d\mu(t)d\nu(g) = \int_M \int_G f(gt)d\nu(g)d\mu(t).$$

By transitivity of G on M and invariance of ν, the inner integral on the right depends on f but not on t. Call it $\bar{\nu}(f)$. Then

$$\nu(1)\mu(f) = \bar{\nu}(f)\mu(1).$$

So that if μ' is another invariant measure on M then

$$\mu(f)\mu'(1) = \mu'(f)\mu(1)$$

□

1.4. REMARKS: **a.** The proof shows also that any right invariant normalized measure on a compact metric group G is equal to any normalized left invariant measure.

b. It is easily checked that the unique normalized invariant measure on G is also invertable invariant i.e. $\int_G f(t)d\mu(t) = \int_G f(t^{-1})d\mu(t)$.

In what follows μ will denote the normalized Haar measure on the space in question so that it may appear twice in the same formula denoting measures on different spaces.

1.5. We pass now to several examples of homogeneous spaces of the group O_n of all $n \times n$ real orthogonal matrices.

EXAMPLES: **a.** $S^{n-1} = \{(x_1, \ldots, x_n) \in \mathbb{R}^n; \sum_{i=1}^n x_i^2 = 1\}$ with either the euclidean or geodesic metric is easily seen to be equivalent to O_n/O_{n-1} as was discussed above (1.2.).

b. *The Stiefel manifolds.* For $1 \leq k \leq n$

$$W_{n,k} = \{e = (e_1, \ldots, e_k)); e_i \in \mathbb{R}^n, (e_i, e_j) = \delta_{ij}, 1 \leq i, j \leq k\}$$

with the metric $\rho(e, f) = (\sum_{i=1}^k d(e_i, f_i)^2)^{\frac{1}{2}}$, d being either the euclidean or the geodesic metric. Note that $W_{n,n} = O_n, W_{n,1} = S^{n-1}$ and $W_{n,n-1} = SO_n = \{T \in O_n;\, det\ T = 1\}$. In general $W_{n,k}$ may be identified with O_n/O_{n-k} via the map $\varphi: O_n \to W_{n,k}$, $\varphi(e_1, \ldots, e_n) = (e_1 \ldots e_k)$.

c. The *Grassman manifolds* $G_{n,k}$, $1 \leq k \leq n$, consists of all k dimensional subspaces of $\mathbb{R}^n$ with the metric being the Hausdorff distance between the unit balls of the two subspaces

$$\rho(\xi, \varsigma) = \sup_{x \in S^{n-1} \cap \varsigma} \rho(x, S^{n-1} \cap \xi) .$$

The equivalence $G_{n,k} = O_n/(O_k \times O_{n-k})$ is again easily verified.

d. If G is any group with invariant metric ρ and G_o is a subgroup we may define a metric d on $M = G/G_o$ by

$$d(t,s) = inf\{\rho(g,h); \varphi(g) = t, \varphi(h) = s\}$$

where $\varphi: G \to M$ is the quotient map. In this way M becomes an homogeneous space of G. Note that in all previous examples the metric given on the homogeneous space of O_n is equivalent, up to a universal constant (not depending on n), to the metric given here.

1.6. The uniqueness of the normalized Haar measure allows us to deduce several interesting consequences.

a. The first remark is that for any $A \subseteq S^{n-1}$ and $x_o \in S^{n-1}$.

$$\mu\{T \in O_n;\ Tx_o \in A\} = \mu(A).$$

b. Next we give two identities. Fix $1 \le k \le n$, for $\xi \in G_{n,k}$ we denote $S(\xi) = S^{n-1} \cap \xi$ the $(k-1)$-dimensional sphere of ξ. Then

$$\int_{S^{n-1}} f d\mu = \int_{G_{n,k}} \int_{S(\xi)} f(t) d\mu_\xi(t) d\mu(\xi)$$

for all $f \in C(S^{n-1})$ where μ_ξ is the normalized Haar measure on $S(\xi)$ (by our convention μ on the left is the normalized Haar measure on S^{n-1} and on the right on $G_{n,k}$.

We identify $\mathbb{R}^{2n}$ with $\mathbb{C}^n$ (by introducing a complex structure in one of the possible ways). For each k we denote the collection of complex k-dimensional subspaces of $\mathbb{C}^n$ by $\mathbb{C}\,Gn,k$ and the unit sphere of any $\xi \in \mathbb{C}\,G_{n,k}$ by $\mathbb{C}\,S(\xi)$ (which can be identified with S^{2k-1}). $\mathbb{C}\,G_{n,k}$ is again an homogeneous space and we get an identity similar to the previous one

$$\int_{S^{2n-1}} f d\mu = \int_{\mathbb{C}\,G_{n,k}} \int_{\mathbb{C}\,S(\xi)} f(t) d\mu_\xi(t) d\mu(\xi).$$

Note that here one integrates on a much smaller space, $\mathbb{C}\,Gn,k$, than the one in the first identity which, adjusted to the dimensions here, would be $G_{2n,2k}$.

2. THE ISOPERIMETRIC INEQUALITY ON S^{n-1} AND SOME CONSEQUENCES

2.1. We begin with the statement of the classical isoperimetric inequality on the sphere. A proof is given in Appendix I. A simple proof of a version of Corollary 2.2. (which is sufficient for the applications here) is given in Appendix V.

For a set A in a metric space (M,ρ) and $\varepsilon > 0$ we denote $A_\varepsilon = \{t; \rho(t,A) \le \varepsilon\}$. In what follows we use the geodesic metric on S^{n-1}.

THEOREM: *For each* $0 < a < 1$ *and* $\varepsilon > 0$, $min\{\mu(A_\varepsilon);\ A \subseteq S^{n-1},\ \mu(A) = a\}$ *exists and is attained on* A_o *- a cap of suitable measure (i.e.,* $A_o = B(x_o, r)$ *for any* $x_o \in S^{n-1}$ *and* r *such that* $\mu(B(r)) = a$*, where* $B(r) = B(x_o,r) = \{x; \rho(x,x_o) \le r\}$*).*

Using this for the case $a = 1/2$ (in which case A_o is half a sphere) we get:

2.2. COROLLARY: *if* $A \subseteq S^{n+1}$ *with* $\mu(A) \ge 1/2$ *then* $\mu(A_\varepsilon) \ge 1 - \sqrt{\pi/8}e^{-\varepsilon^2 n/2}$

PROOF: By theorem 2.1 it is enough to evaluate $\mu(B(\pi/2+\varepsilon))$. Note that $cos^n\theta$ is proportional to the n-volume of S_θ - the set of points on S^{n+1} which are of distance θ from $B(1/2)$ (this is an n-dimensional sphere of radius cos θ). It follows easily (draw a picture) that

$$h(\varepsilon,n) = \mu(B(\pi/2+\varepsilon)) = \int_{-\pi/2}^{\varepsilon} \cos^n\theta d\theta \Big/ \int_{-\pi/2}^{\pi/2} \cos^n\theta d\theta \ .$$

Let $I_n = \int_0^{\pi/2} \cos^n\theta d\theta$ and use the change of variables $\theta = \tau/\sqrt{n}$ and the inequality $\cos t \le e^{-t^2/2}$, $0 \le t \le \pi/2$, to get

$$\begin{aligned}
1 - h(\varepsilon,n) &= \int_\varepsilon^{\pi/2} \cos^n(\theta)d\theta/2I_n = (1/\sqrt{n})\int_{\varepsilon\sqrt{n}}^{(\pi/2)\sqrt{n}} \cos^n(\tau/\sqrt{n})d\tau/2I_n \\
&\le (1/\sqrt{n})\int_{\varepsilon\sqrt{n}}^{(\pi/2)\sqrt{n}} e^{-\tau^2/2}d\tau/2I_n \le (1/\sqrt{n})e^{-\varepsilon^2 n/2}\int_0^{(\pi/2-\varepsilon)\sqrt{n}} e^{-t^2/2}dt/2I_n \\
&\le (1/\sqrt{n})e^{-\varepsilon^2 n/2}\int_0^\infty e^{-t^2/2}dt/2I_n = \frac{e^{-\varepsilon^2 n/2}\sqrt{\pi/2}}{2\sqrt{n}I_n} \ .
\end{aligned}$$

This computation (with $\varepsilon = 0$) also gives $\sqrt{n}\ I_n \le \sqrt{\pi/2}$. To evaluate I_n from below notice that integration by parts gives $I_k = ((k-1)/k)I_{k-2}$ which implies $\sqrt{k}\ I_k \ge \sqrt{k-2}\ I_{k-2}$. Therefore,

$$\sqrt{n}\ I_n \ge min(I_1, \sqrt{2}\ I_2) = min(1, \sqrt{2}\ \pi/4) = 1$$

and

$$1 - h(\varepsilon, n) \leq \sqrt{\pi/8}\; e^{-\varepsilon^2 n/2} .$$

□

REMARK: It can actually be checked that $\sqrt{n}\, I_n \underset{n\to\infty}{\longrightarrow} \sqrt{\pi/2}$.

2.3. The next corollary and theorem are crucial for the applications of the isoperimetric inequality to Banach spaces.

For a continuous function f on S^{n+1} we denote by $\omega_f(\varepsilon)$ its modulus of continuity, $\omega_f(\varepsilon) = \sup\{|f(x) - f(y)|; \rho(x,y) \leq \varepsilon\}$. We denote by M_f the *median* (also called the *Levy mean)* of f, i.e., M_f is a number such that both

$$\mu\{x \in S^{n+1}; f(x) \leq M_f\} \geq 1/2 \text{ and } \mu\{x \in S^{n+1}; f(x) \geq M_f\} \geq 1/2.$$

COROLLARY (Levy's lemma): *Let* $f \in C(S^{n+1})$ *and let* $A = \{x; f(x) = M_f\}$ *then* $\mu(A_\varepsilon) \geq 1 - \sqrt{\pi/2}\; e^{-\varepsilon^2 n/2}$.

PROOF: Note that $A_\varepsilon = (f \leq M_f)_\varepsilon \cap (f \geq M_f)_\varepsilon$ and use 2.2.

□

Notice that the values of f on A_ε are very close to M_f. Indeed, if ε is such that $\omega_f(\varepsilon) \leq \delta$ then $|f(x) - M_f| \leq \delta$ on A_ε. So the content of the previous corollary is that *a well behaved function is "almost" a constant on "almost" all the space.* This phenomenon of *concentration of measure* around one value of the function will appear over and over again in these notes.

2.4. In the next theorem we trade off the set of large measure on which the function is almost constant with a set with linear structure - a subspace of large dimension. The symbol $[\cdot]$ denotes the integer part function.

THEOREM: *For* $\varepsilon, \theta > 0$ *and an integer* n *let* $k(\varepsilon,\theta,n) = [\varepsilon^2 n/(2\log 4/\theta)]$. *Let* $f \in C(S^{n+1})$ *then, for all* $\varepsilon, \theta > 0$, *there exists a subspace* $E \subseteq \mathbb{R}^{n+2}$ *with* $\dim E = k \geq k(\varepsilon,\theta,n)$ *and a* $\theta -$ *net* N in $S(E) = S^{n+1} \cap E$ *such that*

(i) $|f(x) - M_f| \leq \omega_f(\varepsilon)$ *for all* $x \in N$

and

(ii) $|f(x) - M_f| \leq \omega_f(\varepsilon) + \omega_f(\theta)$ *for all* $x \in E \cap S^{n+1}$.

(ii) follows from *(i)*. The proof of *(i)* consists of the following two lemmas. As in 2.3. $A = \{x; f(x) = M_f\}$.

2.5. LEMMA: *For any* $N \subseteq S^{n+1}$ *with* $|N| < \sqrt{\pi/2}\; e^{\varepsilon^2 n/2}$ *there exists a* $T \in O_{n+2}$ *such that* $TN \subseteq A_\varepsilon$. *Consequently, for all* $x \in TN$, $|f(x) - M_f| \leq \omega_f(\varepsilon)$.

PROOF: The lemma follows immediately from 1.6.a: for each $x \in S^{n+1}$ $\mu\{T \in O_{n+2};\ Tx \in A_\varepsilon\} = \mu(A_\varepsilon) \geq 1 - \sqrt{\pi/2}\; e^{-\varepsilon^2 n/2}$. Therefore,

$$\mu\{T \in O_{n+2};\ \ Tx \in A_\varepsilon\ \ for\ every\ x \in N\} \geq 1 - |N|\sqrt{\pi/2}\ e^{-\varepsilon^2 n/2} > 0\ .$$

□

2.6. For later use we state the next lemma in a more general framework.

LEMMA: *For every normed space X with $\dim X = k$ there exists a $\theta - net$ N in $S(X) = \{x \in X; \|x\| = 1\}$ with $|N| \leq (1+2/\theta)^k \leq e^{k\ \log 3/\theta}$.*

PROOF: Let $\{x_i\}_{i=1}^n$ be a maximal set in $S(X)$ with the property that $\|x_i - x_j\| \geq \theta$ for $i \neq j$. Then $\{x_i\}_{i=1}^n$ is an $\theta - net$. The open balls $B(x_i, \theta/2)$ are pairwise disjoint and are all contained in $B(0, 1+\theta/2)$. Comparing the volume of $B(0,1+\theta/2)$ with that of $\cup_{i=1}^n B(x_i,\theta/2)$ we get

$$n \cdot (\theta/2)^k \leq (1+\theta/2)^k$$

or

$$n \leq (1+2/\theta)^k$$

□

2.7. REMARK: An inspection of the proof shows that, taking a bit smaller k in Theorem 2.4. one can get the conclusion for most subspaces of dimension k : For $k \leq [\varepsilon^2 n/(4 log\ 4/\theta)]$ *the measure of the set $E_k \subseteq G_{n+2,k}$ of all k dimensional subspace for which the conclusions of Theorem 2.4 hold satisfies $\mu(E_k) \geq 1 - \sqrt{\pi/2}\ e^{-\varepsilon^2 n/4}$.*

2.8. We now indicate briefly a different way to prove Theorem 2.4.

First we estimate the measure of a cap in S^{k-1} from below

$$\mu(B(\theta)) = \frac{\int_o^\theta \sin^{k-2} t dt}{2\ I_{k-2}} \geq \frac{\int_{\theta/2}^\theta \sin^{k-2} t dt}{2\ I_{k-2}} \geq \frac{\theta \sin^{k-2} \theta/2}{4\ I_{k-2}}\ .$$

Next we use the identity 1.6.b to get that there exists a k-dimensional subspace E such that

$$\mu(E \cap A_\varepsilon) \geq 1 - \sqrt{\pi/2}\ e^{-\varepsilon^2 n/2}\ .$$

If

$$\frac{\theta \sin^{k-2} \theta/2}{4\ I_{k-2}} + 1 - \sqrt{\pi/2}\ e^{-\varepsilon^2 n/2} \geq 1 \qquad (*)$$

then any ball of radius θ in E intersects A_ε and then for any $x \in E$, $|f(x) - M_f| \leq \omega(\varepsilon + \theta)$. Now use (*) to get an estimate on k.

2.9. REMARK: Note that if we want to prove a theorem similar to 2.4. in which the conclusion holds for two functions simultaneously (on the same subspace) we loose very little in the estimate on the dimension. Actually, by Remark 2.7., one can find a subspace E

with dim $E = k \geq [\varepsilon^2 n/(4log\ 4/\theta)]$ for which the conclusion holds for exp $([\varepsilon^2 n/(4log\ 4/\theta)])$ functions simultaneously.

2.10. We conclude this chapter with a few exercises.

a. Let $A \subseteq S^{n-1}$. Then either A_ε or $(A^c)_\varepsilon$ contains $S^{n-1} \cap E$ for some $k(\varepsilon/2, \varepsilon/2, n)$ - dimensional subspace E. ($k(\varepsilon, \theta, n)$ was defined in theorem 2.4.).

b. Fix $k \leq k(\varepsilon/2, \varepsilon/2, n)$, let $A \subseteq S^{n-1}$ be such that for all k-dimensional subspaces E, $A \cap E \neq \emptyset$. Then there exists a k-dimensional subspace E, with $S^{n-1} \cap E \subseteq A_{2\varepsilon}$.

HINT: Consider the function $\rho = \rho(x, A)$ and prove that the median M_ρ is at most ε.

c. For every $A \subseteq S^{n-1}$ we define

$$B_{A,k} = \{E \in G_{n,k}; E \cap A \neq \emptyset\}$$

and

$$I_{A,k} = \{E \in G_{n,k}; S(E) \subseteq A\} .$$

Fix $\varepsilon > 0$ and $k \leq [\varepsilon^2 n/10\ log\varepsilon^{-1}]$. Prove: if $\mu(B_{A,k}) > 2e^{-\varepsilon^2 n/4}$ then $\mu(I_{A_{4\varepsilon},k}) \geq 1 - 2e^{-\varepsilon^2 n/4}$.

HINT: First use Remark 2.7 to show $\mu(A_{2\varepsilon}) \geq 1/2$.

3. FINITE DIMENSIONAL NORMED SPACES, PRELIMINARIES

3.1. Let X, Y be two n-dimensional normed spaces. The *Banach-Mazur distance* between them is defined as

$$d(X,Y) = inf\{\|T\| \cdot \|T^{-1}\|;\ T: X \to Y\ isomorphism\}.$$

Obviously $d(X,Y) \geq 1$ and $d(X,Y) = 1$ if and only if X and Y are isometric. If $d(X,Y) \leq \lambda$ we say that X and Y are $\lambda-$ *isomorphic*. The notion of the distance also has a geometrical interpretation. If $d(X,Y)$ is small then in some sense the two unit balls $B(X) = \{x \in X; \|x\| \leq 1\}$ and $B(Y) = \{y \in Y; \|y\| \leq 1\}$ are close one to the other. More precisely there is a linear transformation φ such that

$$B(X) \subseteq \varphi(B(Y)) \subseteq d(X,Y)B(X).$$

The Banach-Mazur distance satisfies a multiplicative triangle inequality $(d(X,Z) \leq d(X,Y) \cdot d(Y,Z))$. Also $d(X^*, Y^*) = d(X,Y)$ for all X and Y where * denotes the dual space.

In the next few chapters we will consider the space $\mathbb{R}^n$ with two norms on it. One is a general norm $\|\cdot\|$, the other will always be an euclidean norm $|x| = (x,x)^{1/2}$ induced by some inner product $(\cdot,\cdot)$. We denote $D = \{x \in \mathbb{R}^n; |x| \leq 1\}$ and for $E \subseteq \mathbb{R}^n$, $S(E) = \{x \in E; |x| = 1\}$.

Let a, b be such that

3.1.1. $$a^{-1}|x| \leq \|x\| \leq b|x| \text{ for all } x \in \mathbb{R}^n .$$

We may define the norm, dual to $\|\cdot\|$ relative to $(\cdot,\cdot)$, by

$$\|x\|^* = \sup\left\{\frac{|(x,y)|}{\|y\|};\ y \in \mathbb{R}^n \backslash \{0\}\right\}.$$

We get easily that

3.1.2. $$b^{-1}|x| \leq \|x\|^* \leq a|x| \quad \text{for all } x \in \mathbb{R}^n .$$

Indeed, $|x|^2 = (x,x) \leq \|x\|\ \|x\|^* \leq b|x|\ \|x\|^*$, which gives the left side inequality. On the other hand for all $x, y \in \mathbb{R}^n \backslash \{0\}$

$$\frac{|(x,y)|}{\|y\|} \leq \frac{|x| \cdot |y|}{\|y\|} \leq a|x|$$

so that $\|x\|^* = \sup\left\{\frac{|(x,y)|}{\|y\|}\ ;\ y \neq 0\right\} \leq a|x|$.

3.2. The following two theorems deal with the ellipsoid of maximal volume inscribed in the unit ball of a normed space. We recall that a centrally symmetric ellipsoid in $\mathbb{R}^n$ is the body obtained as the image of the unit ball under linear transformation. Equivalently an ellipsoid is the unit ball of any euclidean (i.e. given by an inner product) norm in $\mathbb{R}^n$. The uniqueness of the ellipsoid of maximal volume holds but not trivially. Since we are not going to use it we shall not prove it.

3.3. THEOREM (F. John). *Let $X = (\mathbb{R}^n, \|\cdot\|)$ be an n-dimensional normed space and let D be the ellipsoid of maximal volume inscribed in $B(X)$. Let $|\cdot|$ be the euclidean norm induced by D (i.e. $D = \{x; |x| \le 1\}$). Then*

$$(1/\sqrt{n})|x| \le \|x\| \le |x|.$$

Consequently, $d(X, \ell_2^n) \le \sqrt{n}$ (ℓ_2^n is the n-dimensional canonical Hilbert space, i.e., $\mathbb{R}^n$ with the norm $|(x_1, \dots x_n)| = (\sum_{i=1}^n x_i^2)^{1/2}$).

We only sketch the proof briefly.

Since $D \subseteq B(X)$, $\|x\| \le |x|$. We want to show that $B(X) \subseteq \sqrt{n}\ D$. Applying a linear transformation, we may assume that $D = \{(x_1, \dots, x_n) \in \mathbb{R}^n;\ \sum_{i=1}^n x_i^2 \le 1\}$. If $B(X) \not\subseteq \sqrt{n}\ D$ then there exists a $p \in B(X)$ with $|p| > \sqrt{n}$. Since $B(X)$ is convex, also $K = conv\ D \cup (\pm p) \subseteq B(X)$. We want to show that K contains an ellipsoid of volume larger than the volume of D, in contradiction to the assumption. Without loss of generality we may assume that $p = (d, 0, \dots, 0)$, $d > \sqrt{n}$. The ellipsoid

$$\{x \in \mathbb{R}^n; x_1^2/a^2 + \sum_{i=2}^n x_i^2/b^2 \le 1\}$$

has volume $a \cdot b^{n-1}$. Vol (D) and as long as $a^2/d^2 + (1 - 1/d^2)b^2 \le 1$ it is contained in K (check). The pair $a = d/\sqrt{n}, b = \sqrt{1 - 1/n}\ /\ \sqrt{1 - 1/d^2}$ satisfies this last requirement and $a \cdot b^{n-1} > 1$.

□

3.4. THEOREM (Dvoretzky-Rogers): *Let $X = (R^n, \|\cdot\|)$ be an n-dimensional normed space and let D be the ellipsoid of maximal volume inscribed in $B(X)$. Then there exists a basis $(x_i)_{i=1}^n$ orthonormal with respect to D such that $1 \ge \|x_i\| \ge 2^{-n/(n-i+1)}$, $i = 1, \dots, n-1$.*

PROOF: We choose an orthonormal system inductively in the following way: x_1 is any vector in D with maximal $\|\cdot\|$ -norm (clearly $\|x_1\| = 1$). Given $x_1, \dots, x_i$ we choose x_{i+1} in $(x_1, \dots, x_i)^{\perp} \cap D$ with maximal possible $\|\cdot\|$ -norm. Then for any $x \in span(x_i, \dots, x_n)$ with $|x| \le 1, \|x\| \le \|x_i\|$.

Consider the ellipsoid

$$E = \left\{ \sum_{j=1}^n a_j x_j;\ \frac{\sum_1^{i-1} a_j^2}{a^2} + \frac{\sum_i^n a_j^2}{b^2} \le 1 \right\}.$$

If $\sum_{j=1}^n a_j x_j \in E$ then $\sum_{j=1}^{i-1} a_j x_j \in aD$ and thus $\|\sum_{j=1}^{i-1} a_j x_j\| \le a$. Also

$\sum_{j=i}^{n} a_j x_j \in bD$ and thus $\|\sum_{j=i}^{n} a_j x_j\| \leq b\|x_i\|$. Choosing $a = 1/2$, $b = 1/(2\|x_i\|)$, we get that $E \subseteq B(X)$. On the other hand Vol $(E) = 2^{-(i-1)}(2\|x_i\|)^{-(n-i)}$ Vol (D), so necessarily $2^{-n+1}\|x_i\|^{-(n-i)} \leq 1$ or $\|x_i\| \geq 2^{-(n-1)/(n-i)}$.

□

4. ALMOST EUCLIDEAN SUBSPACES OF A NORMED SPACE.

As in Chapter 3, let $X = (\mathbb{R}^{n+2}, \|\cdot\|)$ be an (n+2)-dimensional normed space and let $|\cdot|$ be an euclidean norm on $\mathbb{R}^{n+2}$ satisfying 3.1.1.

$$a^{-1}|x| \le \|x\| \le b|x|.$$

We are going to apply the results 2.3. and 2.4. to the function $r(x) = \|x\|$ defined on $S^{n+1} = \{x \in \mathbb{R}^{n+2}; |x| = 1\}$. Note that the modulus of continuity of r satisfies $\omega_r(\varepsilon) \le b \cdot \varepsilon$.

4.1. Inspecting the statement of Theorem 2.4., it seems that if we want $r(x)$ to be close to its median M_r we need to choose both ε and θ of order M_r/b. The next lemma shows that, due to the fact that $r(x)$ is the restriction of a convex and homogeneous function, we can do a little better: θ can be chosen independently of M_r/b. This will be important in some of the applications especially in the proof of Dvoretzky's Theorem.

LEMMA: *Assume $|r(x) - M_r| \le b\varepsilon$ for all x in a θ-net N of $E \cap S^{n+1}$ for some subspace E of $\mathbb{R}^{n+2}$ then,*

$$\frac{1-2\theta}{1-\theta}M_r - \frac{b\varepsilon}{1-\theta} \le \|x\| = r(x) \le \frac{1}{1-\theta}M_r + \frac{b\varepsilon}{1-\theta}.$$

for all $x \in E \cap S^{n+1}$.

PROOF: Let $x \in E \cap S^{n+1}$, by successive approximation one can find $\{y_i\}_{i=1}^{\infty}$ in N and $\{\delta_i\}_{i=1}^{\infty}$ in $\mathbb{R}$ with $|\delta_i| \le \theta^{i-1}$ such that $x = y_1 + \sum_{i=2}^{\infty} \delta_i y_i$. It follows that

$$\|x\| \le \|y_i\| + \sum_{i=2}^{\infty} \delta_i \|y_i\| \le \sum_{i=0}^{\infty} \theta^i (M_r + b\varepsilon) = \frac{1}{1-\theta}(M_r + b\varepsilon).$$

To get the lower estimate, let $y \in N$ be such that $|x - y| \le \theta$. Then, by the first part, $\|x - y\| \le \frac{\theta}{1-\theta}(M_r + b\varepsilon)$ and

$$\|x\| \ge \|y\| - \|x - y\| \ge M_r - b\varepsilon - \frac{\theta}{1-\theta}(M_r + b\varepsilon) = \frac{1-2\theta}{1-\theta}M_r - \frac{1}{1-\theta}b\varepsilon.$$

□

4.2. We now apply Theorem 2.4. and the Lemma above to get an estimate on the dimension of a subspace of $\mathbb{R}^{n+2}$ on which the norm $\|\cdot\|$, satisfying 3.1.1., is almost a multiple of the euclidean norm.

THEOREM: *For any $\delta > 0$ there exists a $c(\delta) > 0$ such that for any n and any norm $\|\cdot\|$ on $\mathbb{R}^{n+2}$ satisfying 3.1.1. there exists a subspace E of $\mathbb{R}^{n+2}$ such that $k = \dim E \ge$*

$c(\delta)\cdot n\cdot(M_r/b)^2$ *and* $(1-\delta)\cdot M_r\cdot|x|\le\|x\|\le(1+\delta)\cdot M_r\cdot|x|$ *for* $x\in E$. *In particular* $d(E,\ell_2^k)\le(1+\delta)/(1-\delta)$.

PROOF: Let $\theta>0$ be such that

$$\frac{1}{1-\theta}<1+\frac{\delta}{2}\quad and\quad \frac{1-2\theta}{1-\theta}>1-\frac{\delta}{2},$$

let $\varepsilon'>0$ be such that

$$\frac{1+\varepsilon'}{1-\theta}<1+\delta \text{ and } \frac{1-2\theta-\varepsilon'}{1-\theta}>1-\delta,$$

and let $\varepsilon=\frac{\varepsilon'\cdot M_r}{b}$. By Theorem 2.4., there exists an $E\subseteq R^{n+2}$ with dim $E\ge\left[\frac{1}{2}\frac{\varepsilon^2 n}{\log 4/\theta}\right]=c(\delta)\frac{M_r^2\cdot n}{b^2}$ and a θ-net N in $E\cap S^{n+1}$ such that

$$|\|x\|-M_r|\le b\varepsilon$$

for all $x\in N$. It follows from Lemma 4.1. that, for all $x\in E$,

$$(1-\delta)\cdot M_r\cdot|x|\le\left(\frac{1-2\theta}{1-\theta}M_r-\frac{b\varepsilon}{1-\theta}\right)|x|\le\|x\|\le\left(\frac{1}{1-\theta}M_r+\frac{b\varepsilon}{1-\theta}\right)|x|\le(1+\delta)\cdot M_r\cdot|x|$$

□

REMARKS: **a.** It is easy to see from the proof that the constant $c(\delta)$ satisfies $c(\delta)\ge c\delta^2/\log\frac{1}{\delta}$ for some absolute constant c.

b. Using Remark 2.7. one may show that the collection of all subspaces $E\subseteq G_{n+2,k}$ having the property $d(E,\ell_2^k)\le(1+\delta)/(1-\delta)$ has measure $\ge 1-exp(-c(\delta)\cdot n\cdot M_r^2/b^2)$.

4.3. We turn now to the problem of finding large euclidean sections in a finite dimensional normed space. Theorem 4.2. reduces this problem to the problem of finding euclidean norms for which M_r/b is as large as possible. For a finite dimensional normed space X and for $\varepsilon>0$, we denote by $k(X,\varepsilon)$ the dimension of the largest Hilbert space which is $(1+\varepsilon)$-isormorphic to a subspace of X.

THEOREM: *Let* $X=(\mathbb{R}^n,\|\cdot\|)$ *and assume* $d(X,\ell_2^n)\le d_n$, *then*

$$k(X,\varepsilon)\ge c(\varepsilon)\cdot n/d_n^2$$

where $c(\varepsilon)\ge c\cdot\varepsilon^2/\log\frac{1}{\varepsilon}$, c *an absolute constant.*

PROOF: There exists an euclidean norm on $\mathbb{R}^n$ such that $a\cdot b\le d_n$ and $a^{-1}|x|\le\|x\|\le b|x|$ for $x\in\mathbb{R}^n$. Obviously $M_r\ge min\{\|x\|;|x|=1\}\ge a^{-1}$ so that $M_r/b\ge d_n^{-1}$. Now apply Theorem 4.2.

□

If for a family of spaces $\{X_n\}_{n=1}^{\infty}$ we have $\sup_n d(X_n, \ell_2^n) = M < \infty$ then we get from the Theorem above that $k(X_n, \varepsilon)/n$ is bounded from below by a constant depending on ε and M alone. For this reason it is enough (for the purpose of getting asymptotic estimates) to deal only with $k(X_n, 1)$ (say). We shall denote $k(X, 1)$ by $k(X)$.

Before continuing with the investigation of the quantity M_r/b for some special families of spaces, we would like to gather some facts about the relation between $k(X)$ and $k(X^*)$.

4.4. Applying Theorem 4.2. to the dual norm $r^*(x) = \|x\|^*$, we get from 3.1.2. that if $X = (\mathbb{R}^{n+2}, \|\cdot\|)$ and

$$a^{-1}|x| \leq \|x\| \leq b|x| \qquad x \in \mathbb{R}^{n+2}$$

then there exists a subspace $F \subseteq \mathbb{R}^{n+2}$ with $k^* = dim\ F \geq c(\delta) \cdot n \cdot (M_{r*}/a)^2$ and

$$(1-\delta) \cdot M_{r*} \cdot |x| \leq \|x\|^* \leq (1+\delta) \cdot M_{r*} \cdot |x| \qquad for\ x \in F.$$

Moreover, if $k \leq c(\delta) \cdot n \cdot min\{(M_r/b)^2, (M_{r*}/a)^2\}$ then Remark b in 4.2. implies that one can find a subspace $E \subseteq \mathbb{R}^n$ with $dim\ E = k$ such that simultaneously

$$(1-\delta) \cdot M_r \cdot |x| \leq \|x\| \leq (1+\delta) \cdot M_r \cdot |x|$$

and

$$(1-\delta) \cdot M_{r*} \cdot |x| \leq \|x\|^* \leq (1+\delta) \cdot M_{r*} \cdot |x| \qquad for\ all\ \ x \in E.$$

We summarize these two remarks (specifying a δ) in the following theorem.

THEOREM: *There exists an absolute constant c such that for every normed space $X = (\mathbb{R}^n, \|\cdot\|)$ and euclidean norm $|\cdot|$ on $\mathbb{R}^n$ satisfying*

$$a^{-1}|x| \leq \|x\| \leq b|x| \qquad x \in \mathbb{R}^n$$

we have

(i)

$$k(X) \cdot k(X^*) \geq c \cdot n^2 \cdot \left(\frac{M_r \cdot M_{r*}}{a \cdot b}\right)^2$$

and

(ii) There exists a subspace $E \subseteq \mathbb{R}^n$ with $dim\ E \geq c \cdot n \cdot min\{(M_r/b)^2, (M_{r*}/a)^2\}$ such that

$$\frac{2}{3} \cdot M_r \cdot |x| \leq \|x\| \leq \frac{4}{3} \cdot M_r \cdot |x| \quad and \quad \frac{2}{3} \cdot M_{r*} \cdot |x| \leq \|x\|^* \leq \frac{4}{3} \cdot M_{r*} \cdot |x|, \quad x \in E.$$

4.5. Note that always $M_r \cdot M_{r*} \geq 1$. Indeed, by the definition of M_r and M_{r*}, there exists an $x \in S^{n-1}$ with $\|x\| \leq M_r$ and $\|x\|^* \leq M_{r*}$. Then,

$$M_r \cdot M_{r*} \geq \|x\|^* \cdot \|x\| \geq (x, x) = 1.$$

Next, one may choose $|\cdot|$ in such a manner that $a \cdot b = d(X, \ell_2^n)$, therefore we get from Theorem 4.4.

COROLLARY:

$$k(X) \cdot k(X^*) \geq c \cdot n^2 / d(X, \ell_2^n)^2.$$

In particular,

$$k(X) \cdot k(X^*) \geq c \cdot n \ .$$

The last statement follows from Theorem 3.2.

4.6. The next proposition evaluates the norm of a projection onto the euclidean section one obtained in Theorem 4.4. *(ii)*.

PROPOSITION: *Let $E \subseteq \mathbb{R}^n$ be a subspace satisfying (ii) of Theorem 4.4. (all is needed is $\|x\| \leq \frac{4}{3} \cdot M_r \cdot |x|$, $\|x\|^* \leq \frac{4}{3} \cdot M_{r*} \cdot |x| \quad x \in E$). Let P be the orthogonal (with respect to $|\cdot|$), projection on E. Then*

$$\|P\| \leq \frac{16}{9} \cdot M_r \cdot M_{r*}$$

($\|P\|$ *is the norm of the projection as an operator from* X to X).

PROOF: For any $x \in \mathbb{R}^n$,

$$|Px|^2 = (Px, Px) = (Px, x) \leq \|Px\|^* \cdot \|x\| \leq \frac{4}{3} \cdot M_{r*} \cdot |Px| \cdot \|x\|.$$

Thus,

$$|Px| \leq \frac{4}{3} \cdot M_{r*} \cdot \|x\|$$

and

$$\|Px\| \leq \frac{4}{3} \cdot M_r \cdot |Px| \leq \frac{16}{9} \cdot M_r \cdot M_{r*} \cdot \|x\|.$$

□

4.7. Putting together Theorem 4.4. and Proposition 4.6. we get the inequality

4.7.1. $$k(X) \cdot k(X^*) \geq c \cdot (n \cdot \|P\| / (a \cdot b))^2$$

where P is a projection onto a subspace satisfying *(ii)* of Theorem 4.4. Moreover, choosing the ellipsoid that gives the distance of X to ℓ_2^n, we get

4.7.2. $$k(X) \cdot k(X^*) \geq c \cdot (n \cdot \|P\| / d(X, \ell_2^n))^2 \ .$$

4.8. Let as before, $X = (\mathbb{R}^n, \|\cdot\|)$ be a normed space and let $|\cdot|$ be an euclidean norm on $\mathbb{R}^n$ satisfying 3.1.1., $a^{-1}|x| \leq \|x\| \leq b|x|$. We related above the search for euclidean subspaces of X with the median M_r of $r(x) = \|x\|$ on the sphere $S^{n-1} = \{x; |x| = 1\}$. The next theorem relates this search of subspaces of X, closer to being euclidean then X itself, to the median of $r^*(x) = \|x\|^*$. In the next chapter we shall use this information to obtain good euclidean subspaces and quotient spaces of X.

THEOREM: *Let X satisfy 3.1.1. Then for any $0 < \phi < 1$ there exists a subspace $E \subseteq X$ with*

$$k = dim\ E \geq (1-\phi)n$$

and

$$c\frac{\phi}{M_{r^*}}|x| \leq \|x\| \leq b|x|$$

for all $x \in E$ where $c > 0$ is an absolute constant.

For the proof we need two lemmas the first of which is a different version of 2.2.

4.9. LEMMA: **a.** *Let $A \subseteq S^{n-1}$ with $\mu(A) \geq \frac{1}{2}$ and let $0 < \varepsilon < \frac{\pi}{2}$. Then for some absolute constant $c > 0$,*

$$\mu(A_{\frac{\pi}{2}-\varepsilon}) \geq 1 - c\sqrt{n}\ \sin^{n-2}\varepsilon$$

b. *Let $0 < \delta < \frac{\pi}{2}$ then for some absolute constant c_1,*

$$\mu(B(\delta)) > \frac{c_1}{n}\ sin^{n-1}\delta$$

(recall that $B(\delta) = B(x_0, r), x_0 \in S^{n-1}$).

PROOF: **a.** With the notation of 2.2,

$$\begin{aligned}\mu(A_{\frac{\pi}{2}-\varepsilon}) &\geq \mu(B(\pi-\varepsilon)) = 1 - \mu(B(\varepsilon))\\ &= 1 - \frac{1}{2I_{n-2}}\int_o^\varepsilon \sin^{n-2}\theta d\theta\\ &\geq 1 - \frac{\sqrt{n-2}}{2}\ \sin^{n-2}\varepsilon\ .\end{aligned}$$

b. Put $t = \frac{1}{(n-2)^{3/2}}$ then,

$$\begin{aligned}\mu(B(\delta)) &= \frac{1}{2I_{n-2}}\int_0^\delta \sin^{n-2}\theta d\theta\\ &\geq \frac{1}{2I_{n-2}}\int_{\delta(1-t)}^\delta \sin^{n-2}\theta d\theta\\ &\geq \frac{\sqrt{n-2}}{2}\ \delta t\ \sin^{n-2}(\delta(1-t))\\ &\geq \frac{c}{n}\sin^{n-1}\delta\end{aligned}$$

□

4.10. LEMMA: *Let $X = (I\!R^n, \|x\|)$. Then there exists a set $A \subseteq S^{n-1}$ with $\mu(A) \geq \frac{1}{2}$ and such that for all $0 < \phi < \frac{\pi}{2}$*

$$M_{r^*}\|x\| \geq \sin\phi \geq \frac{2}{\pi}\phi \qquad x \in A_{\frac{\pi}{2}-\phi}\ .$$

PROOF: Take $A = \{y \in S^{n-1};\ \|y\|^* \le M_{r^*}\}$. Then for $x \in A_{\frac{\pi}{2}-\phi}$ we can find a $y \in A$ with

$$dist(x,y) = \frac{\pi}{2} - \phi,$$

that is,

$$(x,y) = \cos(\frac{\pi}{2} - \phi) = \sin\phi.$$

Then

$$\|x\| \ge \frac{(x,y)}{\|y\|^*} \ge \frac{\sin\phi}{M_{r^*}}.$$

□

4.11. PROOF of 4.8. : As usual we may assume that $|\cdot|$ is the standard euclidean norm. It follows from 9.5.1. below that, for some absolute constant $c > 0$, $\sqrt{n}M_{r^*} \ge cr$. Also, in all the relevant applications we actually have $\sqrt{n}M_{r^*} \ge r$. We therefore assume in this proof that $\sqrt{n}M_{r^*} \ge r$ and we thus trivially have the conclusion for $\phi \le \frac{1}{\sqrt{n}}$ (and k=n) and we may assume $\phi \ge \frac{1}{\sqrt{n}}$.

Let $A \subseteq S^{n-1}$ be as in Lemma 4.10. and fix an integer $k = \lambda(n-2)$, $\frac{1}{2} < \lambda < 1$. Then, averaging over all $k+1$ dimensional subspaces of $\mathbb{R}^n$ and using Lemma 4.9. we find a $(k+1)$-dimensional subspaces $E_{k+1} \subseteq \mathbb{R}^n$ with

$$\mu(A_{\frac{\pi}{2}-\varepsilon} \cap S^k) \ge 1 - c\sqrt{n}\ \sin^{n-2}\varepsilon$$

$(S^k = E_{k+1} \cap S^{n-1})$

4.11.1. NOTE: Changing c to another absolute constant the same conclusion holds for a subset of $G_{n,k+1}$ of measure larger than $\frac{1}{2}$. This will be used in the next Chapter.

If δ is such that

4.11.2.
$$\frac{c_1}{k+1}\sin^k\delta + 1 - c\sqrt{n}\ \sin^{n-2}\varepsilon > 1$$

then, by Lemma 4.9., $A_{\frac{\pi}{2}-\varepsilon} \cap S^k$ intersect any $\delta - cap$ in S^k, that is, $A_{\frac{\pi}{2}-\varepsilon} \cap S^k$ is a $\delta - net$ in S^k.

Take $\delta = \varepsilon - \phi$ then, if 4.11.2. is satisfied, $(A_{\frac{\pi}{2}-\varepsilon})_{\varepsilon-\phi} = A_{\frac{\pi}{2}-\phi}$ covers S^k and by Lemma 4.10. we get that E_{k+1} satisfies the conclusion of the theorem.

It remains to evaluate the largest k for which 4.11.2. is satisfied. Rearranging 4.11.2. and taking $(n-2)-th$ root we get that 4.11.2. is satisfied if

$$\sin\varepsilon \le \mu_n \sin^\lambda(\varepsilon - \phi)$$

with $\mu_n < (\frac{c}{n})^{\frac{3}{2n}}$ for some absolute c.

This is equivalent to

$$(\sin\varepsilon)^{1/\lambda} \le \mu_n(\sin\varepsilon\cos\phi - \cos\varepsilon\sin\phi)$$

or

$$\frac{(\sin\varepsilon)^{\frac{1-\lambda}{\lambda}}}{\cos\phi} \le \mu_n(1 - tan\phi cot\varepsilon)$$

or

$$\frac{1-\lambda}{\lambda} log(\sin\varepsilon) - log\cos\phi \le log\mu_n + log(1 - tan\phi cot\varepsilon)$$

Fix $\varepsilon(= \frac{\pi}{4}$, *say*) and notice that

$$log\cos\phi \approx -\phi^2, \ log(1 - tan\phi \ cot\varepsilon) \approx -\phi$$

and

$$|log\mu_n| \le \frac{3}{2n} log\frac{n}{c} << 1/\sqrt{n} \le \phi.$$

Consequently, we get, for small $\phi \ge \frac{1}{\sqrt{n}}$, that 4.11.2. is satisfied as long as

$$\frac{1-\lambda}{\lambda} \ge c\phi \quad (c \ absolute)$$

or, as long as

$$\lambda < 1 - c\phi \quad (c \ absolute)$$

This proves the theorem (use $c\phi$ instead of ϕ).

□

5. ALMOST EUCLIDEAN SUBSPACES OF ℓ_p^n SPACES, OF GENERAL n-DIMENSIONAL NORMED SPACES, AND OF QUOTIENT OF n-DIMENSIONAL SPACES

5.1. In order to use Theorem 4.2. in special spaces one has to evalute M_r/b from below. It turns out that it is much easier to evaluate the average

$$A_r = \int_{S^{n-1}} \|x\| d\mu(x)$$

rather than the median M_r. The next lemma shows that typically the two quantities are close each to the other.

LEMMA: *There exists a constant C such that if 3.1.1. holds with $b \leq \sqrt{n}$ then*

$$|A_r - M_r| < C.$$

If $a \cdot b \leq \sqrt{n}$ than

$$1/2 \leq M_r^{-1} \cdot A_r \leq C$$

PROOF: By 2.3,

$$\mu\{x \in S^{n-1}; \big|\|x\| - M_r\big| > b \cdot t\} \leq \sqrt{\pi/2}\, e^{-t^2 n/2} .$$

Thus, if $b \leq \sqrt{n}$,

$$\mu(\big|\|x\| - M_r\big| > s) \leq \sqrt{\pi/2}\, e^{-s^2/2}$$

and

$$|A_r - M_r| \leq \int_{S^{n-1}} \big|\|x\| - M_r\big| \leq \sqrt{\pi/2} \int_0^\infty e^{-s^2/2} ds = \pi/2.$$

To prove the last claim note that we may assume, multiplying the norm by a constant, that $a \leq 1$, $b \leq \sqrt{n}$. Also $M_r \geq a^{-1} \geq 1$ so that by the first part

$$|A_r - M_r| \leq \pi/2 \leq \pi/2 \cdot M_r$$

and

$$A_r \leq (\pi/2 + 1) M_r.$$

The left hand side inequality $A_r \geq M_r/2$ holds with no restriction on a or b.

□

REMARKS: **1.** The proof also shows that, if $b \leq \sqrt{n}$,

$$|(\int_{S^{n-1}} \|x\|^p)^{1/p} - M_r| \leq C_p$$

for all $1 \leq p < \infty$ with C_p depending on p only.

2. If $a \cdot b = o(\sqrt{n})$ then the proof shows that $M_r^{-1} \cdot A_r = 1 + o(1)$.

5.2. The family of ℓ_p^n spaces, $1 \leq p \leq \infty$, plays an important role in the local theory of normed spaces. ℓ_p^n is the space $\mathbb{R}^n$ equipped with the norm

$$\|(x_1, \ldots, x_n)\|_p = (\sum_{i=1}^{n} |x_i|^p)^{1/p} \qquad p < \infty$$

$$\|(x_1, \ldots, x_n)\|_\infty = \max_{1 \leq i \leq n} |x_i| .$$

As is well known $(\ell_p^n)^* = \ell_q^n$ where $1/p + 1/q = 1$ $(1/\infty = 0)$. We are going to apply now Theorem 4.2. and Corollary 4.5. to the special case of ℓ_p^n spaces.

5.3. *Euclidean sections of* ℓ_1^n

Note that by Holder's inequality

$$\|x\|_2 < \|x\|_1 \leq \sqrt{n}\|x\|_2 \qquad \textit{for all } x \in \mathbb{R}^n.$$

So that using $\|\cdot\|_2$ as the euclidean norm in 4.2. we get that $b = \sqrt{n}$. To evaluate M_r we use Lemma 5.1. and the evaluation of I_n from 2.2.

$$M_r(\ell_1^n) \geq C^{-1} \cdot A_r = C^{-1} \int_{S^{n-1}} \|x\|_1 \, d\mu(x) = C^{-1} \sum_{i=1}^{n} \int_{S^{n-1}} |x_i| d\mu(x) =$$

$$= C^{-1} \cdot n \cdot \int_{S^{n-1}} |x_1| d\mu(x) = C^{-1} \cdot n \cdot \int_{-\pi/2}^{\pi/2} \cos\,\theta (\sin\,\theta)^{n-2} d\theta / I_{n-2}$$

$$\geq C^{-1} \cdot n \cdot \frac{2}{n-1} \cdot \sqrt{2(n-2)/\pi} \geq d \cdot \sqrt{n}$$

for some absolute constant d. Therefore, using Theorem 4.2., we get

$$k(\ell_1^n) \geq c \cdot n \cdot M_r^2 / b^2 \geq \delta \cdot n$$

for some absolute constant δ.

5.4. *Euclidean sections of* $\ell_p^n, 1 < p < \infty$.

We shall need two facts.

FACT 1: *for* $1 < p < \infty$ $\quad d(\ell_p^n, \ell_2^n) \leq n^{|1/p - 1/2|}$

PROOF: By Holder's inequality we get, for $1 \leq p \leq 2$,

$$(\sum_{i=1}^{n} a_i^2)^{1/2} \leq (\sum_{i=1}^{n} |a_i|^p)^{1/p} \leq n^{1/p-1/2} (\sum_{i=1}^{n} a_i^2)^{1/2}$$

For $q > 2$ one gets the result by duality or by a similar inequality.

FACT 2: *For $q > 2$ there exists a constant C_q such that $k(\ell_q^n) \leq C_q \cdot n^{2/q}$ for all n.*

We shall prove Fact 2 in 5.6. below.

Now, corollary 4.5. implies that, for $1 < p < 2$ and $q > 2$ such that $1/p + 1/q = 1$,

$$C_q \cdot n^{2/q} \cdot k(\ell_p^n) \geq k(\ell_q^n) \cdot k(\ell_p^n) \geq c \cdot n^2 / d(\ell_q^n, \ell_2^n)^2 \geq c \cdot n^{1+2/q}.$$

Since clearly $k(\ell_p^n) < n$, we get that we actually have equivalence in all the inequalities above. We conclude with three corollaries.

a. For $q > 2$, $k(\ell_q^n)/n^{2/q}$ is bounded from above and below by constants depending only on q.

b. $n \geq k(\ell_p^n) \geq c_p \cdot n$, $p < 2$, c_p depends only on p

c. $n^{|1/p-1/2|} \geq d(\ell_p^n, \ell_2^n) \geq c_p \cdot n^{|1/p-1/2|}$, $1 < p < \infty$, c_p depends only on p.

We remark in passing that the constant in *b* can actually be choosen to be independent of p (using a proof similar to 5.3.). The constant in c is known to be 1, i.e., $d(\ell_2^n, \ell_q^n) = n^{1/2-1/q}$.

Using Proposition 4.6., one can show that $\ell_q^n, q > 2$, contains a subspace of dimension $\approx n^{2/q}$ which is "nicely" isomorphic to $\ell_2^{n^{2/q}}$ and "nicely" complemented.

5.5. For the proof of fact 2 we need the notion of the Rademacher functions which will play an important role also later in these notes.

Define, for $k = 1, 2, \ldots$ and $0 \leq t \leq 1$

$$r_k(t) = \mathit{sign}\,\sin\,(\pi 2^k t)$$

For example,

$$r_1(t) = \begin{cases} 1 & t < 1/2 \\ -1 & t > 1/2 \end{cases}, \quad r_2(t) = \begin{cases} 1 & t < 1/4, \quad 1/2 < t < 3/4 \\ -1 & 1/4 < t < 1/2, \quad 3/4 < t \end{cases} \quad \ldots$$

This is a sequence of ± 1 valued functions (disregarding a set of measure zero) which is (statistically) independent. Any other sequence with these two properties will do for our purposes. For example, it will sometimes be more convenient for us to use a sequence $(r_i(t))_{i=1}^{\infty}$ defined on the probability space $\{-1,1\}^{\mathbb{N}}$ endowed with the product measure obtained by assigning the measure 1/2 to each of -1 and 1, where $r_i(t)$ is defined, for $t = (t_1, t_2, \ldots)$, by

$$r_i(t) = t_i \qquad i = 1, 2, \ldots .$$

Averages of the form

$$\int_0^1 \|\sum_{i=1}^n r_i(t) x_i\|^p \, dt$$

with x_i in some normed space and $1 \le p < \infty$ will play an essential role in these notes. Note that, in view of the second definition of $(r_i)_{i=1}^{\infty}$, these averages can be represented also as

$$\int_0^1 \|\sum_{i=1}^n r_i(t)x_i\|^p = Ave_{\varepsilon_i=\pm1}\|\sum_{i=1}^n \varepsilon_i x_i\|^p.$$

As an example note that if $x_i \in H, i = 1, \ldots, n$, H a Hilbert space, then the parallelogram equality implies that

$$\int_0^1 \|\sum_{i=1}^n r_i(t)x_i\|^2 = \sum_{i=1}^n \|x_i\|^2.$$

The following inequality regarding linear combinations of the Rademacher functions will be used in the proof of Fact 2 momentarily. It will be used over and over again throughout these notes. We postpone its proof until Chapter 7.

KHINCHINE'S INEQUALITY: *For* $1 \le p < \infty$ *there exist constants* $0 < A_p$, $B_p < \infty$ *such that*

$$A_p(\sum_{i=1}^n |a_i|^2)^{1/2} \le (\int_0^1 |\sum_{i=1}^n a_i r_i|^p \, d(t))^{1/p} \le B_p(\sum_{i=1}^n |a_i|^2)^{1/2}$$

for every n *and every choice of* $a_1, \ldots, a_n$.

The value of the best constants A_p, B_p is known ([Haa]). We remark only (and this will follow from the proof in Chapter 7) that $B_p \approx \sqrt{p}$ while A_p remain bounded away from zero for all p.

5.6. *Proof of fact 2*: Let

$$u_j = (u_{j,1}, \ldots, u_{j,n}), \quad j = 1, \ldots, k = k(\ell_q^n)$$

be vectors in ℓ_q^n such that

$$(\sum_{j=1}^k |a_j|^2)^{1/2} \le \|\sum_{j=1}^k a_j u_j\|_q \le 2(\sum_{j=1}^k |a_j|^2)^{1/2}$$

for any choice of $a_1, \ldots, a_k$. Then, for any $t \in [0, 1]$,

$$k^{q/2} = (\sum_{j=1}^k r_j^2(t))^{q/2} \le \|\sum_{j=1}^k r_j(t)u_j\|_q^q = \sum_{i=1}^n |\sum_{j=1}^k r_j(t)u_{j,i}|^q .$$

Integrating over $[0, 1]$ we get, by Khinchine's inequality,

$$k^{q/2} \le \sum_{i=1}^n \int_0^1 |\sum_{j=1}^k r_i(t)u_{j,i}|^q \, dt \le B_q^q \sum_{i=1}^n (\sum_{j=1}^k u_{j,i}^2)^{q/2}. \qquad (*)$$

Now, for each fixed i

$$\sum_{j=1}^k u_{j,i}^2 \le (\sum_{\ell=1}^n |\sum_{j=1}^k u_{j,i}u_{j,\ell}|^q)^{1/q} = \|\sum_{j=1}^k u_{j,i}u_j\|_q \le 2(\sum_{j=1}^k u_{j,i}^2)^{1/2}.$$

So that

$$(\sum_{j=1}^{k} u_{j,i}^2)^{1/2} \leq 2$$

and from (*)

$$k^{q/2} \leq (2 \cdot B_q)^q \cdot n \ .$$

That is

$$k(\ell_q^n) \leq (2 \cdot B_q)^2 \cdot n^{2/q}$$

□

5.7. *Euclidean sections of* ℓ_∞^n.

The proof of Fact 2 in 5.6. shows also that $k(\ell_\infty^n) \leq C \cdot log\ n$ for some absolute constant C. Indeed, since

$$\|x\|_\infty \leq \|x\|_p \leq n^{1/p} \cdot \|x\|_\infty$$

we get that

$$d(\ell_\infty^n, \ell_{log\ n}^n) \leq e.$$

Consequently, since $B_p \approx \sqrt{p}$,

$$k(\ell_\infty^n) \approx k(\ell_{log\ n}^n) \leq (2 \cdot B_{log\ n})^2 \cdot n^{2/log\ n} \leq C \cdot log\ n.$$

Using Theorem 4.2. we are going to show that $k(\ell_\infty^n) \approx log\ n$. Using the ℓ_2^n norm as the euclidean norm in 4.2. we have that $b = 1$. So that it is enough to prove that $M_r \geq c \cdot (log\ n/n)^{1/2}$. By lemma 5.1. it is enough to get a similar estimate for A_r. For later purpose we prove a more general estimate.

LEMMA: *For* $1 \leq m \leq n$,

$$\int_{S^{n-1}} \max_{1\leq i\leq m} |x_i| d\mu(x) \geq c \cdot \left(\frac{log\ m}{n}\right)^{1/2}$$

for some absolute constant $c > 0$. *In particular, for* $r(x) = \max_{1\leq i\leq n} |x_i| \leq 1$, $A_r \geq c \cdot (log\ n/n)^{1/2}$.

PROOF: Let ν be the measure on $\mathbb{R}^n$ with density $exp(-\pi \cdot \sum_{i=1}^n t_i^2)$. This measure is a probability measure and is clearly invariant under isometries of ℓ_2^n, i.e., under O_n. If $f\colon S^{n-1} \to \mathbb{R}$ is any integrable function, define $\hat{f}(t) = \|t\|_2 \cdot f(\frac{t}{\|t\|_2})$ in $\mathbb{R}^n \backslash \{0\}$. Then $\int_{\mathbb{R}^n} \hat{f}(t) d\nu(t)$ is invariant under the action of O_n and the uniqueness of the Haar measure μ on S^{n-1} implies the existence of a constant λ_n such that

5.7.1. $$\int_{\mathbb{R}^n} \hat{f}(t) d\nu(t) = \lambda_n \int_{S^{n-1}} f(t) d\mu(t).$$

Using polar coordinates and putting $f \equiv 1$ in 5.7.1. one sees that

$$\lambda_n = \int_{\mathbb{R}^n} \|t\| d\nu(t) \leq C \cdot \sqrt{n}$$

for some absolute constant C. This reduces the problem to the problem of estimating $\int_{\mathbb{R}^n} \max_{1\le i\le m} |t_i| d\nu(t)$ from below.

Now, for any $a > 0$,

$$\nu\{\max_{1\le i\le m} |t_i| < \alpha\} = \int_{-\alpha}^{\alpha} \cdots \int_{-\alpha}^{\alpha} exp(-\pi \cdot \sum_{i=1}^{n} t_i^2) dt_1 \dots dt_n =$$

$$= (\int_{-\alpha}^{\alpha} e^{-\pi t^2} dt)^m = (1 - 2\int_{\alpha}^{\infty} e^{-\pi t^2} dt)^m \le (1 - c\cdot e^{-\pi\alpha^2})^m .$$

Choosing $\alpha = \varepsilon \cdot \sqrt{\log m}$ for some absolute ε we get that the last quantity is $\le 1/2$ so that

$$\nu\{\max_{1\le i\le m} |t_i| \ge \varepsilon\sqrt{\log m}\} \ge 1/2$$

and

$$\int_{\mathbb{R}^n} \max_{1\le i\le m} |t_i| d\nu(t) \ge 1/2 \cdot \varepsilon \cdot \sqrt{\log m}.$$

Combining this with 5.7.1. and the estimate for λ_n we get

$$\int_{S^{n-1}} \max_{1\le i\le m} |x_i| d\mu(x) = \lambda_n^{-1} \int_{\mathbb{R}^n} \max_{1\le i\le m} |t_i| d\nu(t) \ge c(\log m/n)^{1/2}$$

□

5.8. We are now in a position to prove Dvoretzky's Theorem.

THEOREM: *There exists an absolute constant c such that $k(X) \ge c \cdot \log n$ for any n-dimensional normed space X.*

PROOF: Let D be the ellipsoid of maximal volume contained in $B(X)$. Let $|\cdot|$ be the norm associated with it. Then by F. John's Theorem 3.2.

$$n^{-1/2}|x| \le \|x\| \le |x| \quad forall \quad x \in X.$$

In particular, $b = 1$ and $a \cdot b \le \sqrt{n}$, so that we may use Lemma 5.1. to evaluate M_r.

By the Dvoretzky-Rogers Lemma 3.3. there is an orthonormal basis $x_1, \dots, x_n$ with $\|x_i\| \ge 1/4$, $i = 1, \dots [n/2]$. Now, for each fixed $t \in [0,1]$,

$$A_r = \int_{S^{n-1}} \|\sum_{i=1}^{n} a_i x_i\| d\mu(a) = \int_{S^{n-1}} \|\sum_{i=1}^{n} r_i(t) a_i x_i\| d\mu(a). \qquad (*)$$

Note that, by the triangle inequality, for all $\{y_i\} \subseteq X$

$$\int_0^1 \|\sum_{i=1}^{n} r_i(t) y_i\| dt = Ave_{\varepsilon_i=\pm1} \|\sum_{i=1}^{n} \varepsilon_i y_i\| \ge \max_{1\le i\le m} \|y_i\|.$$

So, integrating (*) over [0,1] and using Lemma 5.7., we get,

$$A_r = \int_{S^{n-1}} \int_0^1 \|\sum_{i=1}^{n} r_i(t) a_i x_i\| dt d\mu(a) \ge$$

$$\ge \int_{S^{n-1}} \max_{1\le i\le n} \|a_i x_i\| d\mu(a) \ge \frac{1}{4} \int_{S^{n-1}} \max_{1\le i\le [n/2]} |a_i| d\mu(a) \ge c \cdot (\log n/n)^{1/2}$$

for some absolute constant c.

By Lemma 5.1., we get now that $M_r \geq c(\log n/n)^{1/2}$ for some absolute c. Finally, using Theorem 4.2. we get the desired result.

□

5.9. We conclude this chapter with a theorem stating that every n-dimensional normed space has a subspace of a quotient space (= a quotient space of a subspace) of dimension proportional to n which is close to being euclidean.

THEOREM: *Let X be an n-dimensional normed spaces and let $\lambda < 1$. There exists a quotient space Y of a subspace of X with $\dim Y \geq \lambda n$ and*

$$d(Y, \ell_2^{\dim Y}) \leq c(1-\lambda)^{-2}|\log(1-\lambda)|$$

where c is an absolute constant.

PROOF: We shall need one fact which will be proved only in Chapter 15. Namely, there exists an euclidean norm $r(x) = |x|$ on X for which $M_r M_{r^*} \leq c \log(d(X, \ell_2^n))$ for some absolute c (and $a^{-1}|x| \leq \|x\| \leq b|x|$ with $a/M_{r^*} \leq \sqrt{n}$).

By Theorem 4.8. there exists, for any $0 < \phi < 1$, a subspace $E \subseteq X$ with $\dim E = k \geq (1 - c\phi)n$ (c absolute) and

$$\frac{\phi}{M_{r^*}}|x| \leq \|x\| \leq b|x| \qquad x \in E\ .$$

Moreover, (Note 4.11.1.) this holds for a subset of $G_{n,k}$ (the Grassmann manifold) of measure larger than or equal to $\frac{1}{2}$. Using Lemma 5.1. and averaging over $G_{n,k}$ we get that for a subset of $G_{n,k}$ of measure larger than or equal to $\frac{1}{2}$, $M_{E,r}$ (the median of r restricted to $S_E = \{x \in E; |x| = 1\}$) is smaller or equal than $c\, M_r$ for an absolute c. (Actually, for <u>any</u> E with $\dim E = k$, $M_{E,r} \leq c\sqrt{\frac{n}{k}}\, M_r$. This follows from 9.5.1. below. We prefer not be use it here). Consequently, we can find an $E \subseteq X$ with $\dim E = k \geq (1 - c\phi)n$ on which

$$\frac{\phi}{M_{r^*}}|x| \leq \|x\| \leq b|x| \qquad x \in E$$

and

$$M_{E,r} \leq c\, M_r\ .$$

Since

$$b^{-1}|x| \leq \|x\|^* \leq \frac{M_{r^*}}{\phi}|x| \qquad x \in E$$

we can find by the same argument a subspace F of E^* with

$$\dim F \geq (1 - c\phi)^2 n \quad (c \text{ absolute})$$

and

$$\frac{\phi}{M_r}|x| \leq \|x\|^* \leq \frac{M_{r^*}}{\phi}|x|\ .$$

In particular,

$$d(F, \ell_2^{\dim F}) \le \frac{M_r M_{r^*}}{\phi^2} \le c\phi^{-2} \log d(X, \ell_2^n) .$$

Note that F^* is a subspace of a quotient of X. Since $\log d(X, \ell_2^n) \le \log n$ this proves the Theorem up to a $\log n$ factor. To get rid of that $\log n$, define for $t = \frac{k}{n}, k = 1, \ldots, n$,

$$f(t) = \inf\{d(F, \ell_2^{tn});\ F \text{ a subspace of a quotient of } X \text{ with } \dim F = tn\} .$$

Then the proof above shows

$$f(\lambda t) \le c \frac{1}{(1-\sqrt{\lambda})^2} \log f(t)$$

(for any $0 < \lambda < 1$ such that $\lambda t = \frac{k}{n}$ for some $k = 1, \ldots, n$). We may extend f to all of $(\frac{1}{n}, 1]$ in such a way that this inequality still holds. In particular,

5.9.1. $$f(s^2) \le c \frac{1}{(1-\sqrt{s})^2} \log f(s), \quad \frac{1}{n} < s < 1$$

We also have $1 \le f(s) \le n$. The result follows by iterating 5.9.1. We indicate briefly how to perform the iterations.

Put

$$g(s) = f(s) \left(2 \frac{c}{(1-s^{\frac{1}{4}})^2} \log \frac{c}{(1-s^{\frac{1}{4}})^2} \right)^{-1} .$$

Then, for s close enough to 1 (independently of n),

5.9.2. $$g(s^2) \le \log g(s)$$

as long as $g(s) \ge e$ (check).

Iterating 5.9.2. we get

$$g(s^{2^t}) \le \log^{(t)} g(s)$$

where $\log^{(t)} x = \log \log^{(t-1)} x$ and $\log^{(1)} x = \log x$.

Given $0 < \alpha < 1$, let t be the first such that $\log^{(t)} n \le e$ and take s such that $s^{2^t} = \alpha$. Then,

$$g(\alpha) \le \log^{(t)} g(s) \le e$$

and we get the desired result.

□

6. LEVY FAMILIES

6.1. Let (X, ρ, μ) be a metric space (X, ρ) equipped with a Borel probability measure μ. For a subset $A \subseteq X$ and $\varepsilon > 0$ we define, as in Chapter 2,

$$A_\varepsilon = \{x \in X;\ \rho(x, A) \leq \varepsilon\}.$$

Let *diam* X be the diameter of X and assume *diam* $X \geq 1$.

The *concentration function* $\alpha(X, \varepsilon)$ is defined for any $\varepsilon > 0$ by

$$\alpha(X, \varepsilon) = 1 - inf\{\mu(A_\varepsilon);\ A \subseteq X \text{ Borel with } \mu(A) \geq \frac{1}{2}\}$$

DEFINITION: A family (X_n, ρ_n, μ_n), $n = 1, 2, \ldots$ of metric probability spaces is called a Levy family if for every $\varepsilon > 0$,

$$\alpha(X_n, \varepsilon diam X_n) \underset{n \to \infty}{\longrightarrow} 0.$$

The family is called a *normal Levy family* with constants c_1, c_2, if

$$\alpha(X_n, \varepsilon) \leq c_1 exp(-c_2 \varepsilon^2 n).$$

Note that any normal Levy family is a Levy family (since we assume $diam X_n \geq 1$). The omission of the factor $diam X_n$ in the definition of a normal Levy family comes because that way most of the examples below become Levy families with their natural metric and natural enumeration.

As we have already seen in specific examples in Chapter 2, in a Levy family we necessarily have the phenomenon of concentration of measure around one value of a function. If $f : X \to \mathbb{R}$ is a function with modulus of continuity $\omega_f(\varepsilon)$ and with median M_f then we get

6.1.1. $$\mu(|f - M_f| \leq \omega_f(\varepsilon)) \geq 1 - 2\alpha(X, \varepsilon).$$

Indeed, put $A = \{x \in X;\ f(x) \leq M_f\}$, $B = \{x \in X;\ f(x) \geq M_f\}$ then $\mu(A), \mu(B) \geq 1/2$ and

$$\begin{aligned}\mu(|f - M_f| \leq \omega_f(\varepsilon)) &= \mu[(f \leq M_f + \omega_f(\varepsilon)) \cap (f \geq M_f - \omega_f(\varepsilon))] \\ &\geq \mu(A_\varepsilon \cap B_\varepsilon) \geq 1 - 2\alpha(X, \varepsilon)\ .\end{aligned}$$

That is, f is concentrated close to M_f on most (in the sense of measure) of X, if $\alpha(X, \varepsilon)$ is small enough.

In 2.2. we have shown that $\{S^{n+1}\}_{n=1}^\infty$, with the natural metric and probability measure, is a normal Levy family with constants $c_1 = \sqrt{\pi/8}$ and $c_2 = 1/2$. We applied this fact in

Chapters 2 and 4. In this Chapter we shall give a lot of other examples of Levy families. We shall see that a number of natural families of metric probability spaces are Levy families. Most of these examples found applications in the Local Theory of normed spaces. A few of them play a central role in this theory and for them we present a full detailed proof here or in later Chapters; the rest are only briefly sketched.

6.2. Let $E_2^n = \{-1,1\}^n$ with the (product) measure

$$P_n(A) = |A| \cdot 2^{-n}, \ A \subseteq E_2^n$$

and with the normalized Hamming metric

$$d_n(x,y) = \frac{1}{n}|\{i; x_i \neq y_i\}| = \frac{1}{2n}\sum_{i=1}^{n}|x_i - y_i| \ , \qquad x,y \in E_2^n.$$

THEOREM: *(E_2^n, d_n, P_n) is a normal Levy family with constants $c_1 = 1/2$ and $c_2 = 2$.*

The theorem follows from an isoperimetric inequality (i.e. identifying the sets $A \subseteq E_2^n$ for which the infimum $inf\{P(A_\varepsilon); P(A) \geq 1/2\}$ is attained) of [Har]. We shall present a different proof (giving somewhat worse constants) in Chapter 7 below. The same proof will prove also the next result of [Ma].

6.3. Let Π_n be the group of all permutations of the set $\{1,\ldots,n\}$. Let P_n be the normalized counting measure;

$$P_n(A) = |A|/n!, \qquad A \subseteq \Pi_n$$

and let d_n be the normalized Hamming metric;

$$d_n(\pi,\rho) = \frac{1}{n}|\{i; \pi(i) \neq \rho(i)\}|, \quad \pi,\rho \in \Pi_n$$

THEOREM: *(Π_n, d_n, P_n) is a normal Levy family with constants $c_1 = 2$, $c_2 = 1/64$.*

6.4. The following few examples of Levy families are consequences of a general isoperimetric inequality for connected riemannian manifolds due to Gromov [Gr1]. Appendix I, written by Gromov, contains the proof together with the necessary definitions.

Let μ_X be the normalized riemannian volume element on a connected riemannian manifold without boundary X and let $R(X)$ be the Ricci curvature of X.

THEOREM: *Let $A \subseteq X$ be measureable and let $\varepsilon > 0$ then*

$$\mu_X(A_\varepsilon) \geq \mu_X(B_\varepsilon)$$

where B is a ball on the sphere $r \cdot S^n$ with $n = dimX$, and r such that

$$R(X) = R(r \cdot S^n)(= (n-1)/r^2)$$

and $\mu_X(A) = \mu(B)$, μ being the normalized Haar measure on $r \cdot S^n$.

The value of $R(X)$, known in some examples ([C.E.]), together with the computation for the measure of a cap in 2.2. leads to the following examples.

6.5.1. THEOREM: *The family $SO_n = \{T \in O_n;\ detT = 1\}, n = 1,2\ldots,$ with the metric discribed in 1.5.b and the normalized Haar measure is a normal Levy family with constants* $c_1 = \sqrt{\pi/8}, c_2 = 1/8$.

6.5.2. Similarly for each m the family $X_n = S^n \times S^n \times \ldots S^n(m - times), n = 1,2,\ldots,$ with the product measure and the metric

$$d(x,y) = (\sum_{i=1}^{m} \rho(x_i, y_i)^2)^{1/2}, x = (x_1, \ldots, x_m), y = (y_1, \ldots, y_m) \in X_n$$

(ρ-the geodesic metric in S^n), is a normal Levy family with constants $c_1 = \sqrt{\pi/8}$, $c_2 = 1/2$.

6.6. Next we show that homogeneous spaces of SO_n inherit the property of being Levy family. Let G be a subgroup of SO_n and let $V = SO_n/G$. Let μ be the Haar measure on V and let d_n be the metric introduced in 1.5.d, i.e.

$$d_n(t,s) = inf\{\rho(g,h); \varphi g = t, \varphi h = s\}$$

where φ is the quotient map.

Clearly $\mu(A \subseteq V) = \mu(\varphi^{-1}(A) \subseteq SO_n)$. By the definition of d_n, $\varphi^{-1}(A_\epsilon) \supseteq (\varphi^{-1}A)_\epsilon$. Therefore, if $\mu(A \subseteq V) \geq 1/2$, then $\mu(\varphi^{-1}(A) \subseteq SO_n) \geq 1/2$ and $\mu(A_\epsilon) \geq \mu((\varphi^{-1}A)_\epsilon)$. We conclude

THEOREM: *Let, for $n = 1,2,\ldots, G_n$ be a subgroup of SO_n with the metric described above and with the normalized Haar measure μ_n. Then $(SO_n/G_n, d_n, \mu_n), n = 1,2,\ldots,$ is a normal Levy family with constants $c_1 = \sqrt{\pi/8}$ and $c_2 = 1/8$.*

6.7. Theorem 6.6. together with Examples 1.5.b and 1.5.c implies immediately that the following families are normal Levy families with constants $c_1 = \sqrt{\pi/8}$, $c_2 = 1/8$:

6.7.1. Any family of Stiefel manifolds $\{W_{n,k_n}\}_{n=1}^{\infty}$ with $1 \leq k_n \leq n,\ n = 1,2,\ldots,$

6.7.2. Any family of Grassman manifolds $\{G_{n,k_n}\}_{n=1}^{\infty}$ with $1 \leq k_n \leq n,\ n = 1,2,\ldots.$

6.7.3. Exercise: Define a natural metric and probability measure on $V_{n,k} = \{\xi \in G_{n,k}; x \in S(\xi)\}$ and show that it is a normal Levy family (here, as before, $S(\xi)$ is the unit euclidean sphere of the subspace ξ).

6.8. We consider in this section a *compact connected riemannian manifold* M with μ being the *normalized riemannian volume element* of M. Then the *Laplacian* $-\Delta$ on M has its spectrum consisting of eigenvalues $0 = \lambda_0 < \lambda_1(M) \leq \lambda_2(M) \ldots$. The first non zero eigenvalue λ_1 may be represented as the largest constant such that

$$\lambda_1 \|f\|_{L_2}^2 \leq (-\Delta f, f) = \int_M |\nabla f|^2 d\mu \qquad (*)$$

for every "sufficiently smooth" function f on M such that $\int_M f = 0$. We refer the reader to [B.G.M.] for more details. For the reader who is not familiar or does not feel comfortable with the notions above, we give a somewhat more detailed explanation on a model situation. We hope that this degression will help the reader to develop some intuition.

Let X be a smooth n-dimensional compact connected C^2-smooth submanifold without boundary of the euclidean space $(\mathbb{R}^N, |\cdot|)$. Then norm $|\cdot|$ induces a metric on X. We, however, consider a different metric, called the *length* metric, ρ: for $x, y \in X, \rho(x,y)$ is the length (with respect to $|\cdot|$ of the shortest curve in X joining x and y. The Lebesgue measure in $\mathbb{R}^n$ induces a measure on X. We normalize it so that the measure of X equals 1 and denote the normalized measure by μ. We now have a metric probability space (X, ρ, μ)

For every $x \in X$ let T_x denote the *tanget* (n-dimensional) *plane* to X at x. We write $T_x = x + \xi_x$ where ξ_x is the parallel plane through the origin. Similarly the *normal plane* (of dimension $N - n$) is $N_x = x + \varsigma_x$ where $\varsigma_x \in G_{N,N-n}$ is the orthogonal complement of ξ_x in $\mathbb{R}^N$. The space of pairs

$$T(X) = \{(x,y);\quad x \in X,\quad y \in T_x\}$$

is called the *tangent bundle* of X. We shall also use the *dual bundle*

$$T^*(X) = \{(x,z);\quad x \in X,\quad z \in T_x^* = x + \xi_x^*\}$$

where ξ_x^* is the dual space to ξ_x (which we do not identify with ξ_x).

For $\varepsilon > 0$ let $U_\varepsilon(X)$ denote an $(|\cdot|-)$ ε- neighbourhood of X in $\mathbb{R}^N$. There exists an $\varepsilon > 0$ such that for every $y \in U_\varepsilon(X)$ there exists a unique $x \in X$ with $y \in N_x$ (see [Mi]). Therefore, given any $f \in C(X)$ we may extend it to $\tilde{f} \in C(U_\varepsilon(X))$ by $\tilde{f}(y) = f(x)$ for $y \in N_x$. If $\tilde{f}$ is smooth enough we may consider $\Delta\tilde{f}$ (where standardly in $\mathbb{R}^N$, $\Delta = \sum_{i=1}^N \frac{\partial^2}{\partial x_i^2}$). Note that the derivatives of $\tilde{f}$ in the directions of N_x are zero and therefore Δ may be written in terms of derivatives in the direction of T_x. For example, if $X = S^n \subseteq \mathbb{R}^{n+1}$, then, rewritting Δ in spherical coordinates, we obtain the Laplacian on S^n, (see [Vi], p. 493).

It is sometimes more usuful to see the Laplacian as constructed in two steps. In the first step we introduce the gradient operation ∇. For every Lipschitz function f on X the gradient is defined for almost all x and for such x is an element of $\xi_x, \nabla f(x) = \nabla f_{|x} \in \xi_x$. Indeed, it is more convenient to work with the dual space ξ_x^* and to consider $\nabla f(x)$ as a linear functional (an element of ξ_x^*) we denote it then by $df(x)(\in \xi_x^*)$ of course $|df(x)|^* = |\nabla f(x)|$. Then $df: X \to T^*(X)$ or alternatively $df: T(X) \to \mathbb{R}$ by $df(x,y) = (df(x), y)$ for $y \in \xi_x$. The operator d from a dense subspace of $L_2(X)$ into $L_2(X \to T^*(X), \mu)$ (the space of vector valued functions $F: X \to T^*(X)$ with $\|F\| = (\int (|F(x)|_x^*)^2 d\mu)^{1/2}, |\cdot|_x^*$being the norm in ξ_x^*) can be extended to become a closed operator (also denoted by d)with the domain $D(d)$. Let d^* be the adjoint operator, then the Laplacian is given by

$$-\Delta = d^* d: D(d) \subseteq L_2(X,\mu) \to L_2(X,\mu).$$

By construction, $-\Delta$ is non-negative. It is known that for a compact connected manifold $-\Delta$ has a discrete spectrum, $\{\lambda_i\}_{i=0}^{\infty}$, $\lambda_i \geq 0$, and that $-\Delta f = 0$ only for $f \equiv$ const. It follows that $\lambda_0 = 0$ and $\lambda_1 > 0$. The min-max principle implies now the inequality $(*)$.

6.9. THEOREM: *Let (M, ρ, μ) be a compact connected riemannian manifold (μ the normalized riemannian volume). Let $A \subseteq M$ with $a = \mu(A) > 0$. Then, for all $\varepsilon > 0$,*

$$\mu(A_\varepsilon) \geq 1 - (1 - a^2) exp(-\varepsilon\sqrt{\lambda_1} log(1 + a)).$$

PROOF: Let A, B be two open subsets of M, $\mu(A) = a, \mu(B) = b$ and $\rho(A, B) = \rho > 0$. Consider the function

$$f(x) = 1/a - (1/\rho)(1/a + 1/b) min(\rho(x, A), \rho), \quad x \in M.$$

Let $\int_M f d\mu = \alpha$ and apply $(*)$ of 6.8. to get

$$\lambda_1 \|f - \alpha\|_{L_2}^2 \leq \int |\nabla f|^2 d\mu.$$

Now, f is a constant on each of A and B so that the integration is on a set of measure $= 1 - a - b$. f is a Lipschitz function with Lipchitz constant $\leq (1/\rho)(1/a + 1/b)$ so that $|\nabla f| \leq (1/\rho)(1/a + 1/b)$. Consequently we get

$$\lambda_1 \|f - \alpha\|_{L_2}^2 \leq (1/\rho^2)(1/a + 1/b)^2 (1 - a - b).$$

On the other hand

$$\begin{aligned} \lambda_1 \|f - \alpha\|_{L_2}^2 &\geq \lambda_1 \Big(\int_A (f - \alpha)^2 d\mu + \int_B (f - \alpha)^2 d\mu \Big) \\ &= \lambda_1 ((1/a - \alpha)^2 \cdot a + (1/b + \alpha)^2 \cdot b) \\ &\geq \lambda_1 (1/a + 1/b). \end{aligned}$$

Therefore,

$$\lambda_1 \cdot \rho^2 \leq (1/a + 1/b)(1 - a - b) \leq \frac{1 - a - b}{a \cdot b}$$

or

$$b \leq \frac{1 - a}{1 + \lambda_1 \rho^2 a}.$$

Fix a $\delta > 0$ and consider the sequence of pairs $(A_i, B_i), i = 0, 1, \ldots$, of subsets of M defined inductively by:

$$A_0 = A, \quad B_0 = ((A_0)_\delta)^c$$

$$A_{i+1} = (A_i)_\delta, \quad B_{i+1} = ((A_{i+1})_\delta)^c, \qquad i = 0, 1 \ldots .$$

Let $a_i = \mu(A_i)$, $b_i = \mu(B_i)$, then

$$b_i \leq \frac{1 - a_i}{1 + \lambda_1 \cdot \delta^2 \cdot a_i}$$

Since $a_i \geq a$ for all i,

$$b_i = 1 - a_{i+1} \leq \frac{1 - a_i}{1 + \lambda_1 \cdot \delta^2 \cdot a}$$

Take $\delta = 1/\sqrt{\lambda_1}$, then

$$1 - a_{i+1} \leq \frac{1 - a_i}{1 + a}$$

and, by induction

$$1 - a_i \leq \frac{1 - a}{(1 + a)^i}.$$

If $\varepsilon = i \cdot \delta$ for some i, then $i = \varepsilon\sqrt{\lambda_1}$ and

$$1 - \mu(A_\varepsilon) = 1 - a_i \leq (1 - a)exp(-\varepsilon\sqrt{\lambda_1}log(1 + a)).$$

In the general case pick the i such that $i\delta \leq \varepsilon < (i + 1)\delta$. Then

$$\begin{aligned} 1 - \mu(A_\varepsilon) &\leq (1 - \mu(A_{i \cdot \delta})) \leq (1 - a)exp(-\varepsilon\sqrt{\lambda_1}\, log(1 + a) + log(1 + a)) \\ &\leq (1 - a^2)exp(-\varepsilon\sqrt{\lambda_1}log(1 + a)). \end{aligned}$$

EXAMPLES: **1.** $\lambda_1(S^n) = n$ (this is the so called Wirtinger's inequality see [Gr2] and [Vi] p. 494) and we get: if $A \subseteq S^n$, $\mu(A) \geq 1/2$ then $\mu(A_\varepsilon) \geq 1 - c_1 e^{-c_2\varepsilon\sqrt{n}}$. This is somewhat weaker than what we got in 2.2.

2. Let $\mathbb{T}^n = \Pi_{i=1}^n S^1$ then $\lambda_1(\mathbb{T}^n) = 1$ (on $\mathbb{T}^1$ the eigenfunctions of $-\Delta$ are $\cos kx, \sin kx, k = 0, 1, \ldots$ with eigenvalues $\{k^2\}_{k=0}^\infty$. For $\mathbb{T}^n$ the eigenfunctions are product of functions of the form $\cos kx, \sin kx$ in the different variables; clearly one cannot get a positive eigenvalue smaller than 1).

Here we get $\mu(A) = 1/2 \Rightarrow \mu(A_\varepsilon) \geq 1 - c_1 exp(-c_2\varepsilon)$. Again, a better result may be obtained using 6.5.2., however not in a straight forward way. More precisely,

There exist constants $c_1, c_2 > 0$ such that for every Borel subset A of $\mathbb{T}^n$

$$\mu(A) = \frac{1}{2} \Rightarrow \mu(A_\varepsilon) \geq 1 - c_1 exp(-c_2\varepsilon^2)$$

Note that in this example *diam* $\mathbb{T}^n \approx \sqrt{n}$ and so that $\varepsilon = \delta\sqrt{n}$ is small with respect to the diameter and for such ε we get a significant estimate.

7. MARTINGALES

This section contains applications of some martingale inequalities to the local theory of Banach spaces. We begin with some elementary definitions.

7.1. Let $(\Omega, \mathcal{F}, P)$ be a probability space, let $\mathcal{G}$ be a sub σ-algebra of $\mathcal{F}$ and let $f \in L_1(\Omega, \mathcal{F}, P)$. Then $\mu(A) = \int_A f dP, A \in \mathcal{G}$, defines a measure on $\mathcal{G}$ which is absolutely continuous with respect to $P|\mathcal{G}$. Consequently, by the Radon- Nikodym Theorem, there exists a unique $h \in L_1(\Omega, \mathcal{G}, P)$ such that $\int_A h dP = \int_A f dP$ for all $A \in \mathcal{G}$. We call this h the *conditional expectation* of f with respect to $\mathcal{G}$ and denote $h = E(f|\mathcal{G})$.

The operator $f \to E(f|\mathcal{G})$ is easily seen to be a linear positive operator of norm one on all the L_p spaces, $1 \le p \le \infty$.

7.2. Some additional properties of the conditional expectation operator are:

7.2.1. if $\mathcal{G}'$ is a sub σ-algebra of $\mathcal{G}$ then $E(E(f|\mathcal{G})|\mathcal{G}') = E(f|\mathcal{G}')$.

7.2.2. if $g \in L_\infty(\Omega, \mathcal{G}, P)$ then $E(f \cdot g|\mathcal{G}) = g \cdot E(f|\mathcal{G})$.

7.2.3. if $\mathcal{G}$ is the trivial σ-algebra, $\mathcal{G} = \{\phi, \Omega\}$, then $E(f|\mathcal{G})$ is the expectation of f

$$E(f|\mathcal{G}) = Ef = \int f dP.$$

7.3. Given a sequence $\mathcal{F}_1 \subseteq \mathcal{F}_2 \subseteq \ldots \subseteq \mathcal{F}$ of σ-algebras, a sequence $f_1, f_2, \ldots$ of functions $f_i \in L_1(\Omega, \mathcal{F}_i, P)$ is said to be a *martingale* with respect to $\{\mathcal{F}_i\}_{i=1}^\infty$ if $E(f_i|\mathcal{F}_{i-1}) = f_{i-1}$ for $i = 2, 3, \ldots$

A reader who feels uncomfortable with these definitions is advised to look at the following special case which is typical for most of the applications. Ω is a finite set, P is the normalized counting measure $P(A) = |A|/|\Omega|$. $\{\Omega_i\}_{i=1}^k$ is a sequence of partitions of Ω each of which refines the previous one. $\mathcal{F}_i$ is the algebra generated by Ω_i. For a function f on $\Omega, E(f|\mathcal{F}_i)$ is simply the function which is constant on atoms of Ω_i, the constant on each atom is the averaged value of f on this atom.

7.4. All the martingale inequalities in these notes have a common feature: they evaluate from above the probability of a large deviation of a function from its expectation.

LEMMA: *Let* $f \in L_\infty(\Omega, \mathcal{F}, P), \{\phi, \Omega\} = \mathcal{F}_0 \subseteq \mathcal{F}_1 \subseteq \ldots \subseteq \mathcal{F}_n = \mathcal{F}$, and *put* $d_i = E(f|\mathcal{F}_i) - E(f|\mathcal{F}_{i-1}),\ i = 1, \ldots n$. *Then, for all* $c \ge 0$, $P(|f - Ef| \ge c) \le 2exp(-c^2/\ 4\sum_{i=1}^n \|d_i\|_\infty^2)$.

PROOF: Notice that $\{E(f|\mathcal{F}_i)\}_{i=0}^n$ is a martingale with respect to $\{F_i\}_{i=0}^n$ so that $E(d_i|\mathcal{F}_{i-1}) = 0,\ i = 1, \ldots n$. Using the numerical inequality

$$e^x \le x + e^{x^2}, \quad x \in \mathbb{R}$$

we get, for all $\lambda \in \mathbb{R}$,

$$E(e^{\lambda d_i}|\mathcal{F}_{i-1}) \leq E(\lambda d_i|\mathcal{F}_{i-1}) + E(e^{\lambda^2 d_i^2}|\mathcal{F}_{i-1}) \leq e^{\lambda^2\|d_i\|_\infty^2}.$$

It follows from the properties of conditional expectation that for all $1 \leq i \leq n$,

$$Eexp(\lambda \sum_{j=1}^{i} d_j) = E(E(exp(\lambda \sum_{j=1}^{i} d_j)|\mathcal{F}_{i-1})) = E(exp(\lambda \sum_{j=1}^{i-1} d_j) \cdot E(exp\lambda d_i|\mathcal{F}_{i-1})$$
$$\leq E\ exp(\lambda \sum_{j=1}^{i-1} d_j) \cdot exp(\lambda^2\|d_i\|_\infty^2).$$

Now, for all $\lambda > 0$,

$$P(f - Ef \geq c) = P(\sum_{j=1}^{n} d_j \geq c) = P(exp(\lambda \sum_{j=1}^{n} d_j - \lambda c) \geq 1)$$
$$\leq E\ exp(\lambda \sum_{j=1}^{n} d_j - \lambda c) \leq exp(\lambda^2 \cdot \sum_{j=1}^{n} \|d_j\|_\infty^2 - \lambda c).$$

Putting $\lambda = c/\ 2\sum_{j=1}^{n}\|d_j\|_\infty^2$, we get

$$P(f - Ef \geq c) \leq exp(-c^2/4\sum_{j=1}^{n}\|d_j\|_\infty^2).$$

Similarly,

$$P(Ef - f \geq c) \leq exp(-c^2/4\sum_{j=1}^{n}\|d_j\|_\infty^2)$$

and

$$P(|f - Ef| \geq c) \leq P(f - Ef \geq c) + P(Ef - f \geq c) \leq 2exp(-c^2/4\sum_{j=1}^{n}\|d_j\|_\infty^2).$$

□

7.5. To illustrate the way one uses the lemma, we will prove the following theorem, already stated in 6.4.

THEOREM: *The family* Π_n *of permutation of* $\{1, \ldots n\}$ *with the metric* $d(\pi, \xi) = \frac{1}{n}|\{i; \pi(i) \neq \xi(i)\}|$ *and the uniform measure* P *(assigning mass* $\frac{1}{n!}$ *to each permutation) is a normal Levy family with constants* $c_1 = 2$ *and* $c_2 = 1/64$.

Later on we will state a more general theorem (7.8.)

7.6. PROOF: Let $\mathcal{F}_j$, $0 \leq j \leq n$, be the algebra of subsets of Π_n generated by the atoms $\{A_{i_1,\ldots,i_j}\ ;\ 1 \leq i_1, \ldots, i_j \leq n\ distinct\}$ where $A_{i_1,\ldots,i_j} = \{\pi; \pi(1) = i_1, \ldots, \pi(j) = i_j\}$.

Then $\{\phi, \Pi_n\} = \mathcal{F}_0 \subseteq \mathcal{F}_1 \subseteq \ldots \subseteq \mathcal{F}_n = 2^{\Pi_n}$. This sequence of algebras has the following crucial property:

Given any j, an atom of $\mathcal{F}_j$, $A = A_{i_1,\dots,i_j}$ and two atoms of F_{j+1}, $B = A_{i_1,\dots,i_j,r}$ and $C = A_{i_1,\dots,i_j,s}$ contained in A; one can find a one to one map $\varphi: B \to C$ such that $d(b,\varphi(b)) \le \frac{2}{n}$ for all $b \in B$. (*)

Assume (*). If f is a function on Π_n with Lipschitz constant 1 and $(f_j)_{j=0}^n$ is the martingale generated by f (i.e. $f_j = E(f|\mathcal{F}_j)$) then, for any A,B,C as in (*), f_{j+1} is a constant on B and C and $|f_{j+1|B} - f_{j+1|C}| \le \frac{2}{n}$. Indeed, $f_{j+1|B} = \frac{1}{|B|}\sum_{\pi\in B} f(\pi)$ and $f_{j+1|C} = \frac{1}{|C|}\sum_{\pi\in C} f(\pi) = \frac{1}{|B|}\sum_{\pi\in B} f(\varphi(\pi))$ so that

$$\left|f_{j+1|B} - f_{j+1|C}\right| \le \frac{1}{|B|}\sum_{\pi\in B}|f(\pi) - f(\varphi(\pi))| \le \frac{1}{|B|}\sum_{\pi\in B} d(\pi,\varphi(\pi)) \le \frac{2}{n}.$$

It follows that for any A, B as in (*) $|f_{j+1|B} - f_{j|A}| \le \frac{2}{n}$. Indeed, $f_{j|A} = \frac{1}{n-j}\sum_{C\subseteq A} f_{j+1|C}$ so that $|f_{j+1|B} - f_{j|A}| \le \frac{1}{n-j}\sum_{C\subseteq A}|f_{j+1|B} - f_{j+1|C}| \le \frac{2}{n}$.

Since this holds for all A, B as in (*) we get that $\|d_{j+1}\|_\infty \le \frac{2}{n}$ for $i = 0,\dots,n-1$. Applying the lemma we get that if f is a function with Lipschitz constant 1 then

$$P(|f - Ef| \ge c) \le 2\ exp(\frac{-c^2 n}{16}). \tag{**}$$

If $A \subseteq \Pi_n$ then $d(\cdot,A)$ is a Lipschitz function with constant one so we can use (**). Choose c such that $2exp(-\frac{c^2 n}{16}) = \frac{1}{2}$, i.e., $c = 4\sqrt{(log\ 4)/n}$. Then

$$P(|d(\cdot,A) - Ed(\cdot,A)| < 4\sqrt{(log\ 4)/n}\ ) > \frac{1}{2}$$

On the other hand if $P(A) \ge \frac{1}{2}$ then $P(d(\cdot,A) = 0) \ge \frac{1}{2}$. So there exists a $\pi \in \Pi_n$ such that

$$|d(\pi,A) - Ed(\cdot,A)| < 4\sqrt{(log\ 4)/n}$$

and $d(\pi,A) = 0$, i.e., $Ed(\cdot,A) < 4\sqrt{(log\ 4)/n}$.

Plugging this back into (**) we get

$$P(d(\cdot,A) \ge c + 4\sqrt{(log\ 4)/n}\ ) \le 2\ exp(-c^2 n/16)$$

or, for all $\varepsilon > 8\sqrt{(log\ 4)/n}$,

$$P(A_\varepsilon^c) = P(d(\cdot,A) > \varepsilon) \le 2\ exp(\frac{-\varepsilon^2 n}{64})$$

It is easily checked that this holds also for $\varepsilon \le 8\sqrt{(log\ 4)/n}$.

To prove (*) let ρ be the permutation which changes r with s and leaves the rest at place. Define $\varphi(\pi) = \rho\circ\pi$ then $\varphi(\pi)(i) = \pi(i)$ for all i except possibly for $i = j+1$ (where $\pi(i) = r, \varphi(\pi)(i) = s$) and for $i = \pi^{-1}(s)$ (where $\pi(i) = s$ and $\varphi(\pi)(i) = \rho\circ\pi(i) = \rho(s) = r$), so $d(\pi,\varphi(\pi)) \le \frac{2}{n}$. □

7.7. The property $(*)$ of Π_n suggests the following definition:

DEFINITION: *Given a finite metric space (Ω, d). We say that (Ω, d) is of length at most ℓ if there exist positive numbers $a_1, \dots, a_n$ with $\ell = (\sum_{i=1}^n a_i^2)^{1/2}$ and a sequence $\{\Omega^k\}_{k=0}^n$, $\Omega^k = \{A_i^k\}_{i=1}^{m_k}$, of partitions of Ω with the following four properties:*

7.7.1. $m_0 = 1$ *i.e.* $\Omega^0 = \{\Omega\}$

7.7.2. $m_n = |\Omega|$ *i.e.* $\Omega^n = \{\{x\}\}_{x \in \Omega}$

7.7.3. *Ω^k is a refinement of Ω^{k-1}, $k = 1, \dots, n$*

7.7.4. *for all $k = 1, \dots, n$, $r = 1, \dots, m_{k-1}$ and i, j such that $A_i^k, A_j^k \subseteq A_r^{k-1}$ there exists a one to one and onto function $\varphi: A_i^k \to A_j^k$ with $d(x, \varphi(x)) \leq a_k$.*

Note that ℓ is always smaller or equal to the diameter of Ω. The proof of the theorem below is a direct generalization of the proof of the previous theorem and we leave it to the reader.

7.8. THEOREM: *Let (Ω, d) be a finite metric space of length at most ℓ, let P be the normalized counting measure.*

(i) Let $f: \Omega \to R$ be a function satisfying $|f(x) - f(y)| \leq d(x, y)$ for all $x, y \in \Omega$. Then for all $c \geq 0$

$$P\{|f - Ef| \geq c\} \leq 2\, exp(-\frac{c^2}{4\ell^2})$$

(ii) Let $B \subseteq \Omega, P(B) \geq \frac{1}{2}$, then for all $c \geq 0$

$$P(B_c) \geq 1 - 2\, exp(-\frac{c^2}{16\ell^2}).$$

7.9. Another family of probability metric spaces which can be shown to be a normal Levy family by the same approach is the sequence $E_2^n = \{-1, 1\}^n$, $n = 1, 2, \dots$, with the normalized Hamming metric $d((\varepsilon_i)_{i=1}^n, (\delta_i)_{i=1}^n) = \frac{1}{n}|\{i; \varepsilon_i \neq \delta_i\}|$ and the normalized counting measure. With the natural choice of partitions one gets that the length of $\{-1, 1\}^n$ is at most $1/\sqrt{n}$ so that by 7.8. for all $A \subseteq \{0, 1\}^n$ with $P(A) \geq \frac{1}{2}$ and all $\varepsilon \geq 0$,

$$P(A_\varepsilon) \geq 1 - 2\, exp(-\frac{\varepsilon^2 n}{16}).$$

This proves 6.2. (with different constants).

7.10. REMARK: Note the difference in the order of deduction between this Chapter and Chapter 2. There we used an inequality similar to the one in Theorem 7.8. *(ii)* to deduce an inequality similar to the one in Theorem 7.8. *(i)*. Here the order is reversed. As in Chapters 2 and 4 the inequalities we use in most applications to Banach space theory are for estimating large deviations of nice functions, like the inequalities in Lemma 7.4. and Theorem 7.8. *(i)*.

7.11. Our next theorem is a further abstractization of the method of the proof (7.6.) that Π_n is a Levy family.

Given a compact metric group G with a translation invariant metric d (i.e. $d(g,h) = d(rg,rh) = d(gr,hr)$ for all $g,h,r \in G$) and a closed subgroup H. One can define a natural metric $\bar{d}$ on G/H by

$$\bar{d}(rH,sH) = d(r,sH) = d(s^{-1}r,H).$$

The translation invariance of d implies that this is actually a metric and that $d(r,sH)$ does not depend on the representative r of rH.

7.12. THEOREM: *Let G be a group, compact with respect to a translation invariant metric d. Let $G = G_0 \supseteq G_1 \supseteq \ldots \supseteq G_n = \{1\}$ be a decreasing sequence of closed subgroups of G. Let a_k be the diameter of $G_{k-1}/G_k, k = 1,\ldots,n$. Then*

(i) If $f: G \to R$ is a function satisfying $|f(x) - f(y)| \leq d(x,y)$ for all $x,y \in G$, then for all $c > 0$,

$$\mu(|f - Ef| \geq c) \leq 2\, exp(-c^2/4 \sum_{k=1}^{n} a_k^2)$$

(ii) If $B \subseteq G$ with $\mu(B) \geq \frac{1}{2}$, then, for all $c \geq 0$,

$$\mu(B_c) \geq 1 - 2\, exp(-c^2/16 \sum_{k=1}^{n} a_k^2)$$

(μ is the normalized Haar measure).

PROOF: The implication *(i)* $\Rightarrow$ *(ii)* is obtained as before using the function $f(x) = d(x,B)$.

To prove *(i)*, let $\mathcal{F}_k, k = 0,1,\ldots,n$, be the σ-algebra generated by the sets $\{gG_k\}_{g\in G}$. Note that if $gG_{k-1} \supseteq hG_k$ then $g^{-1}h \in G_{k-1}$.

If both $gG_{k-1} \supseteq h_1G_k$ and $gG_{k-1} \supseteq h_2G_k$, let $s \in G_{k-1}$ be such that $diam(G_{k-1}/G_k) = d(g^{-1}h_1, g^{-1}h_2G_k) = d(g^{-1}h_1, g^{-1}h_2s)$ and define $\varphi: h_1G_k \to h_2G_k$ by

$$\varphi(h_1r) = h_2sr.$$

Then

$$d(h_1r, \varphi(h_1r)) = d(h_1, h_2s) = diam(G_{k-1}/G_k) = a_k.$$

If follows that if we define

$$f_k = E(f|\mathcal{F}_k)$$

then the oscillation of f_k on each atom of $\mathcal{F}_{k-1}$ is at most a_k, so that $\|d_k\|_\infty \leq a_k$ and using Lemma 7.4. we get

$$P(|f - Ef| \geq c) \leq 2\, exp\left\{\frac{-c^2}{4\sum_{k=1}^{n} a_k^2}\right\}$$

□

7.13. The two previous examples (Π_n and $\{-1,1\}^n$) are instances of this theorem. Another example is $\mathbb{T}^n$ with the normalized product measure μ and the ℓ_1 metric

$$d(\bar{t},\bar{s}) = \sum_{i=1}^{n} |t_i - s_i|.$$

Taking $\mathbb{T}^k, k = 0,\ldots,n$, as subgroups, one gets $a_k \le 1$ for all k. Then one can apply the theorem to get e.g. that if $B \subseteq \mathbb{T}^n$ with $\mu(B) \ge \frac{1}{2}$ then

$$\mu(B_c) \ge 1 - 2\, exp(\frac{-c^2}{4n})$$

this should be compared with 6.5. where we used the euclidean distance.

7.14. We bring now two more applications of the method developed here. The first is a proof of Khinchine's inequality used in 5.5. Recall that $\{r_i\}_{i=1}^{\infty}$ are the Rademacher functions.

THEOREM: *For all $1 \le p < \infty$ there exist constants $0 < A_p, B_p < \infty$ such that*

$$A_p^{-1}(\sum_{i=1}^{n} |a_i|^2)^{1/2} \le \|\sum_{i=1}^{n} a_i r_i\|_p \le B_p(\sum_{i=1}^{n} |a_i|^2)^{1/2}$$

for all $\{a_i\}_{i=1}^{n} \subseteq I\!R$. Moreover $B_p/\sqrt{p}$ is bounded and A_p is bounded away from zero.

PROOF: Fix $\{a_i\}_{i=1}^{n} \subseteq \mathbb{R}$ with $\sum_{i=1}^{n} a_i^2 = 1$. $\{\sum_{i=1}^{k} a_i r_i\}_{k=1}^{n}$ is a martingale (with respect to the algebras $\mathcal{F}_k$ = the algebra generated by $r_1,\ldots,r_k$). Therefore,

$$\|d_i\|_\infty = \|a_i r_i\|_\infty = |a_i| \ ,$$

so by Lemma 7.4., for all $c \ge 0$,

$$P(|\sum a_i r_i| > c) \le 2\, exp(-\frac{c^2}{4}).$$

for $1 \le p < \infty$ we get, using integration by parts,

$$\int_0^1 |\sum a_i r_i|^p\, dt = \int_0^\infty t^p\, dP(|\sum a_i r_i| \le t) = p\int_0^\infty t^{p-1} P(|\sum a_i r_i| > t)\, dt$$
$$\le 2p \int_0^\infty t^{p-1} exp(\frac{-t^2}{4}) dt \stackrel{\text{def}}{=} B_p^p < \infty.$$

So, for $2 \le p < \infty$,

$$1 = (\int_0^1 |\sum a_i r_i|^2)^{1/2} \le (\int_0^1 |\sum a_i r_i|^p)^{1/p} \le B_p.$$

The inequality $\frac{t^m}{m!} \le e^t$ easily implies that $\sup_{p\ge 2} B_p/\sqrt{p} < \infty$.

For $p = 1$, let $\theta = \frac{1}{3}$ then $\frac{1}{2} = \frac{\theta}{1} + \frac{1-\theta}{4}$ and, by Holder's inequality,

$$1 = (\int_0^1 |\sum a_i r_i|^2)^{1/2} \le (\int_0^1 |\sum a_i r_i|)^{\theta} (\int_0^1 |\sum a_i r_i|^4)^{(1-\theta)/4}$$
$$\le (\int_0^1 |\sum a_i r_i|)^{\theta} \cdot B_4^{1-\theta}.$$

Consequently,

$$\int_0^1 |\sum a_i r_i| \geq B_4^{-2}.$$

Finally, for $1 \leq p < 2$,

$$B_4^{-2} \leq \int_0^1 |\sum a_i r_i| \leq (\int_0^1 |\sum a_i r_i|^p)^{1/p} \leq (\int_0^1 |\sum a_i r_i|^2)^{1/2} = 1$$

□

7.15. The second application is a different version of 5.3. in which one loses some of the precision in the embedding of $\ell_2^{\alpha n}$ into ℓ_1^n but gains some additional information on the form of the embedding.

THEOREM: *There exist constants* $0 < a,\ b < \infty$ *and* $0 < \alpha < 1$ *such that for all integers* k, n *with* $k \leq \alpha n$ *one can find signs* $\{\varepsilon_{ij}\}_{i=1}^{k}{}_{,}{}_{j=1}^{n}$ *such that*

$$a(\sum_{i=1}^{n} a_i^2)^{1/2} \leq \|\sum_{i=1}^{k} a_i x_i\|_{\ell_1^n} \leq b(\sum_{i=1}^{n} a_i^2)^{1/2}$$

for all $\{a_i\}_{i=1}^{n} \subseteq I\!R$, *where*

$$x_i = \frac{1}{n}\sum_{j=1}^{n} \varepsilon_{ij} e_j \in \ell_1^n, \qquad i = 1, \ldots, k.$$

PROOF: For $\varepsilon = (\varepsilon_{ij})_{i=1}^{k}{}_{,}{}_{j=1}^{n} \in \{-1.1\}^{kn}$ let

$$x_i(\varepsilon) = \frac{1}{n}\sum_{j=1}^{n} \varepsilon_{ij} e_j \in \ell_1^n.$$

Fix $\overline{a} = (a_i)_{i=1}^{k} \subseteq \mathrm{I\!R}$ with $\sum_{i=1}^{k} a_i^2 = 1$ and let

$$f(\varepsilon) = f_{\overline{a}}(\varepsilon) = \|\sum_{i=1}^{k} a_i x_i(\varepsilon)\|_{\ell_1^n}.$$

The triangle inequality implies that for any $\varepsilon, \delta \in \{-1, 1\}^{kn}$

$$|f(\varepsilon) - f(\delta)| \leq \frac{1}{n}\sum_{i=1}^{k} |a_i| \sum_{j=1}^{n} |\varepsilon_{ij} - \delta_{ij}|$$

We want to make f into a Lipschitz function of constant one. So we define a distance function d on $\{-1, 1\}^{kn}$ by

$$d(\varepsilon, \delta) = \frac{1}{n}\sum_{i=1}^{k} |a_i| \sum_{j=1}^{n} |\varepsilon_{ij} - \delta_{ij}|.$$

Using the natural sequence of (kn) σ-algebras, one checks that the length of $(\{-1,1\}^{kn}, d)$ is at most $\sqrt{1/n}$, so, by Theorem 7.8. *(i)*

$$P(|f - Ef| \geq c) \leq 2\ exp(\frac{-c^2 n}{4})$$

P being the normalized counting measure on $\{-1,1\}^{kn}$.

Next we use Khinchine's inequality to evaluate Ef,

$$Ef = E\frac{1}{n}\sum_{j=1}^{n}|\sum_{i=1}^{k} a_i\varepsilon_{ij}| = E|\sum_{i=1}^{k} a_i r_i(t)|.$$

So

$$A_1^{-1} \leq Ef \leq 1$$

and we get that for all $c \geq 0$,

$$P(A_1^{-1} - c \leq f_{\overline{a}} \leq 1 + c) \geq 1 - 2\ exp(\frac{-c^2 n}{4}).$$

Let now N be an θ-net in S^{k-1} with $|N| \leq exp\frac{2k}{\theta}$. Then

$$P\{A_1^{-1} - c \leq f_{\overline{a}} \leq 1 + c,\ for\ all\ \overline{a} \in N\} \geq 1 - 2\ exp(\frac{2k}{\theta} - \frac{c^2 n}{4}).$$

If $A_1 - c \leq f_{\overline{a}} \leq 1 + c$ for all $\overline{a} \in N$. Then, for any $\overline{a} \in S^{k-1}$, we find by successive approximation $\{\overline{u}_i\}_{i=1}^{\infty} \subseteq N$ and $\{\delta_i\}_{i=2}^{\infty}$ with $0 \leq \delta_i < \theta^{i-1}$ such that

$$\overline{a} = \overline{u}_1 + \sum_{i=2}^{\infty} \delta_i \overline{u}_i.$$

Consequently,

$$f_{\overline{a}} \leq (1 + c)\sum_{i=0}^{\infty}\theta^i = \frac{1+c}{1-\theta}.$$

Also, for any $\overline{a} \in S^{n-1}$, find $\overline{b} \in N$ with $|\overline{a} - \overline{b}| < \theta$. Then

$$f_{\overline{a}} \geq f_{\overline{b}} - f_{\overline{a}-\overline{b}} \geq A_1^{-1} - c - \frac{1-c}{1-\theta}\theta,$$

and we get

$$P(A_1^{-1} - c - \frac{1-c}{1-\theta}\theta \leq f_{\overline{a}} \leq \frac{1+c}{1-\theta}\ for\ all\ \overline{a} \in S^{k-1}) \geq 1 - 2\ exp(\frac{2k}{\theta} - \frac{c^2 n}{4}).$$

Given $\xi > 0$ one can find $c > 0$ and $\theta > 0$ such that

$$\frac{1+c}{1-\theta} < 1 + \xi \quad and \quad A_1^{-1} - c - \frac{1-c}{1-\theta}\theta > A_1^{-1} - \xi.$$

Then, for $k < \frac{\theta}{2}(\frac{c^2 n}{2} - log 2)$,

$$1 - 2exp(\frac{2k}{\theta} - \frac{c^2 n}{4}) > 0$$

and one can find $\varepsilon \in \{-1, 1\}^{kn}$ such that

$$A_1^{-1} - \xi \leq f_{\overline{a}}(\varepsilon) \leq 1 + \xi \quad \textit{for all} \quad \overline{a} \in S^{k-1} \ .$$

□

7.16. REMARK: It is known ([Sz1], [Haa]) that $A_1 = \sqrt{2}$ so that one can choose in the Theorem $b = 1 + \xi, a = \sqrt{1/2} - \xi$ for any $\xi > 0$. In that case α seems to depend on ξ and tends to zero as ξ tends to zero.

8. EMBEDDING ℓ_p^m INTO ℓ_1^n

This chapter is devoted to a more complicated application of the martingale method developed in Chapter 7. The proofs and the style here are more technical. The reader who is interested mainly in the outline of the local theory of normed spaces is advised to skip to the next chapter in first reading.

In this chapter we shall estimate the dimension m of ℓ_p^m, $1 < p < 2$, which embed nicely into ℓ_1^n and see that m can be chosen to be proportional to n. This is the first application, in these notes, of the method developed here to normed spaces which does not involve euclidean spaces. In Chapter 10 we shall see some more applications which do not involve euclidean spaces.

8.1. p-Stable Random Variables.

The proof of the main theorem in this chapter uses the notion of p-stable random variables. Since these random variables are an important tool in the local theory of normed spaces, we take this opportunity to discuss some of their properties. Proofs of facts about p-stable variables which are not proved here (8.1.1 and 8.1.2 below) may be found e.g. in [Lo].

DEFINITION: *A random variable g on a probability space $(\Omega, \mathcal{F}, P)$ is called symmetric p-stable, for some $0 < p \leq 2$, if*

$$Ee^{itg} = \int_\Omega e^{it\cdot g(w)} dP(w) = e^{-c|t|^p}$$

for some $c > 0$ and every $-\infty < t < \infty$.

We shall need two facts about these random variables:

8.1.1. There actually are such variables for each $0 < p \leq 2$ (for $p > 2$ there are no such variables!) and

8.1.2. The tail distribution of a symmetric p-stable variable satisfies the inequality

$$P(|g| \geq t) \leq Ct^{-p}\ , \quad t > 0\ ,$$

with C depending on c and p only.

It follows easily that a symmetric p-stable variable belongs to $L_r(\Omega)$ for all $r < p$ (but not for $r = p$).

p-stables can be used to isometrically embed ℓ_p into L_r for $1 \leq r < p \leq \infty$. Indeed, if $g_1, g_2, \ldots$ is a sequence of independent symmetric p-stables with the same distribution (i.e., the c in the Definition 8.1 is the same for all the g_i-s) and $\{a_i\}_{i=1}^n$ are scalars, then,

$$E\ exp\left(it\sum_{j=1}^n a_j g_j\right) = \prod_{i=1}^n E\ exp(ita_j g_j) = \prod_{i=1}^n exp(-c|t|^p|a_j|^p) = E\ exp\left(-c|t|^p\sum_{i=1}^n |a_j|^p\right)\ .$$

So that, if $\Sigma_{j=1}^{n}|a_j|^p = 1$, then $\Sigma_{j=1}^{n} a_j g_j$ and g_1 have the same distribution (we recall that the distribution of a random variable X is determined by its characteristic function Ee^{itX}). Consequently,

$$\left\| \sum_{j=1}^{n} a_j g_j \right\|_r = \left(\sum_{j=1}^{n} |a_j|^p \right)^{1/p} \|g_1\|_r$$

for all n and all scalars $\{a_j\}_{j=1}^{n}$, i.e., the span of the $\{g_i\}_{i=1}^{\infty}$ in L_r is isometric to ℓ_p.

From this it follows (using some technical perturbation arguments) that for any $\varepsilon > 0$ and k there exists an $n = n(k, \varepsilon, r, p)$ such that ℓ_r^n contains a k dimensional subspace E with $d(E, \ell_p^k) \leq 1 + \varepsilon$. The estimate one gets on n this way is very poor. We are going to show that actually n can be chosen to be of order k. To avoid complicated expressions and side-tracked technical computations we restrict our attention only to the case $r = 1$. We begin with a technical approximation lemma permitting us to replace the p-stables by discrete random variables.

8.2. Discretization of p-stable variables

LEMMA: *Let $1 < p < 2$, $\varepsilon > 0$. Let g be a symmetric p-stable random variable on $[0,1]$, endowed with the Lebesgue measure λ, and assume $\|g\|_1 = 1$. Let $g^*(t)$ be the decreasing rearrangement of $|g|$ (i.e., g^* is decreasing and $P(g^* > t) = P(|g| > t)$ for all $0 < t < \infty$) and let $a_i = g^*(\frac{i}{n})$, $i = 1, \ldots, n$. Then there exists an $\alpha = \alpha(\varepsilon, p)$ such that for all $m, n \in \mathbb{N}$ with $m \leq \alpha n$, if $y_1, \ldots, y_m$ is a sequence of independent, symmetric random variables with each of $|y_i|$, $i = 1, \ldots, m$, having the same distribution as that of*

$$y = \sum_{i=1}^{n} a_i \chi_{(\frac{i-1}{n}, \frac{i}{n}]} ,$$

then

$$(1-\varepsilon)\left(\sum_{j=1}^{m} |b_j|^p \right)^{1/p} \leq \left\| \sum_{j=1}^{m} b_j y_j \right\|_1 \leq (1+\varepsilon)\left(\sum_{j=1}^{m} |b_j|^p \right)^{1/p}$$

for all scalars $b_1, \ldots, b_m$. (χ_A is the indicator function of the set A).

PROOF: Draw a picture to check that

$$\|g^* - y\|_1 = \int_0^1 (g^* - y)\, d\lambda \leq \int_0^{1/n} g^* d\lambda .$$

Let $C = C(p)$ be such that

$$P(g^* > t) = P(|g| > t) \leq C \cdot t^{-p} .$$

Then

$$g^*(t) \leq C^{1/p} t^{-1/p}$$

and

$$\int |g^* - y| d\lambda \le \int_0^{1/n} g^* d\lambda \le C^{1/p} \cdot \int_0^{1/n} t^{-1/p} d\lambda = C^{1/p} \frac{p}{p-1} n^{(1-p)/p} .$$

Let $g_1, \ldots, g_m$ be independent symmetric p-stable variables with $\|g_j\|_1 = 1$, $j = 1, \ldots, m$. Put

$$z_j = \Sigma a_i sign\ g_j \chi_{\{a_i \le |g_j| < a_{i-1}\}} , \quad j = 1, \ldots, m .$$

Then, $z_1, \ldots, z_m$ have the same joint distribution as $y_1, \ldots, y_m$ and

$$\|g_j - z_j\| \le C^{1/p} \cdot \frac{p}{p-1} \cdot n^{(1-p)/p} .$$

Consequently, for all scalars $b_1, \ldots, b_n$,

$$\left\| \sum_{j=1}^m b_j (g_j - z_j) \right\|_1 \le C^{1/p} \cdot \frac{p}{p-1} \cdot n^{(1-p)/p} \cdot \sum_{j=1}^m |b_j|$$

$$\le C^{1/p} \cdot \frac{p}{p-1} \cdot \left(\frac{m}{n}\right)^{(p-1)/p} \cdot \left(\sum_{j=1}^m |b_j|^p \right)^{1/p} .$$

Choose α such that $C^{1/p} \cdot \frac{p}{p-1} \cdot \alpha^{(p-1)/p} < \varepsilon$. Then

$$\left\| \sum_{j=1}^m b_j (g_j - z_j) \right\| \le \varepsilon (\Sigma |b_j|^p)^{1/p}$$

and

$$\left\| \sum_{j=1}^m b_j z_j \right\| \le \left\| \sum_{j=1}^m b_j g_j \right\| + \|\Sigma b_j (g_j - z_j)\|$$

$$\le (1+\varepsilon) \left(\sum_{j=1}^m |b_j|^p \right)^{1/p} .$$

Similarly,

$$\left\| \sum_{j=1}^m b_j z_j \right\| \ge (1-\varepsilon) \left(\sum_{j=1}^m |b_j|^p \right)^{1/p} .$$

□

From this lemma it follows immediately that $\ell_p^m \overset{1+\varepsilon}{\hookrightarrow} \ell_1^n$ provided $n \approx m^m$. As we have said above and will show below, this estimate can be greatly improved.

8.3. Weak ℓ_p norms

A second ingredient in the proof of the main result of this chapter (Theorem 8.8) is a martingale inequality (Lemma 8.4) which is stated in terms of the weak ℓ_p norm. For $0 < p < \infty$ and $\overline{\alpha} = (\alpha_1, \ldots, \alpha_k)$ define

$$\|\overline{\alpha}\|_{p,\infty} = \max_{1 \le i \le k} \alpha_i^* \cdot i^{1/p}$$

where $(\alpha_i^*)_{i=1}^k$ is the decreasing rearrangement of $(|\alpha_i|)_{i=1}^k$. The weak ℓ_p norm $\|\cdot\|_{p,\infty}$, satisfies an "upper p-estimate for disjoint vectors", i.e. if $\overline{\alpha}_1,\ldots,\overline{\alpha}_\ell$ are vectors in $\mathbb{R}^k$ such that at each coordinate at most one of them is non-zero, then

$$\left\|\sum_{i=1}^{\ell}\overline{\alpha}_i\right\|_{p,\infty}^p \leq \sum_{i=1}^{\ell}\|\overline{\alpha}_i\|_{p,\infty}^p .$$

Indeed, it is clearly enough to prove this for two such vectors $\overline{\alpha}=\Sigma_{i=1}^k\alpha_ie_i$, $\overline{\beta}=\Sigma_{i=1}^k\beta_ie_i$ $(|\alpha_i|\wedge|\beta_i|=0)$. If $\|\overline{\alpha}+\overline{\beta}\|_{p,\infty}^p=|\alpha_{i_0}|^p\cdot j_0$ then $|\alpha_{i_0}|$ is the j_0 largest among $\{|\alpha_i|\}\cup\{|\beta_i|\}$. Assume $|\alpha_{i_0}|$ is the i-th largest among $\{|\alpha_i|\}$. Then $\|\overline{\alpha}\|_{p,\infty}^p\geq|\alpha_{i_0}|^p\cdot i$, and among $\{|\beta_i|\}$ there are at least j_0-i members larger than or equal to $|\alpha_i|$. So $\|\overline{\beta}\|_{p,\infty}^p\geq|\alpha_{i_0}|^p\cdot(j_0-i)$ and

$$\|\overline{\alpha}\|_{p,\infty}^p+\|\overline{\beta}\|_{p,\infty}^p\geq|\alpha_{i_0}|^p\cdot j_0=\|\overline{\alpha}+\overline{\beta}\|_{p,\infty}^p .$$

8.4. We return now to the martingale technique used in Chapter 7. The next lemma is a consequence of Lemma 7.4.

LEMMA: *Let $(\Omega,\mathcal{F},P)$ be a probability space and let*

$$\{\phi,\mathcal{F}\}=\mathcal{F}_0\subseteq\mathcal{F}_1\subseteq\ldots\subseteq\mathcal{F}_k=\mathcal{F}$$

be a sequence of σ-algebras. Let $f\in L_\infty(\Omega,\mathcal{F},P)$ and put

$$d_i=E(f|\mathcal{F}_i)-E(f|\mathcal{F}_{i-1}) , \quad i=1,\ldots,k .$$

Then, for all $1<p<2$ and all $c>0$,

$$P(|f-Ef|\geq c)\leq 2\,exp\left(\frac{-\delta_pc^q}{\|\{\|d_i\|_\infty\}_{i=1}^k\|_{p,\infty}^q}\right)$$

where $\frac{1}{p}+\frac{1}{q}=1$ and $\delta_p=\frac{(2-p)}{8p(q+1)^q}$.

PROOF: Assume, without loss of generality, that

$$\|(\|d_k\|_\infty)_{k=1}^n\|_{p,\infty}=1$$

and choose a permutation π of $\{1,\ldots,n\}$ so that

$$\|d_{\pi(k)}\|_\infty=\|d_k\|^* , \quad k=1,2,\ldots,n .$$

Thus, we have for $k=1,2,\ldots,n$,

$$\|d_{\pi(k)}\|_\infty\leq k^{-1/p} .$$

Given an integer $N\leq n$ we have:

$$P\left[|\sum_{k=1}^n d_k|\geq(q+1)N^{1/q}\right]\leq P\left[|\sum_{k=1}^N d_{\pi(k)}|\geq qN^{1/q}\right]+P\left[|\sum_{k=N+1}^n d_{\pi(k)}|\geq N^{1/q}\right] .$$

But

$$\left|\sum_{k=1}^{N} d_{\pi(k)}\right| \le \sum_{k=1}^{N} \|d_{\pi(k)}\|_\infty \le \sum_{1}^{N} k^{-1/p} < qN^{1/q}$$

so we get, by Lemma 7.4,

$$\begin{aligned} P\left(|\sum_{k=1}^{n} d_k| \ge (q+1)N^{1/q}\right) &\le 2\ exp\left[\frac{-N^{2/q}}{4\sum_{k=N+1}^{n}\|d_{\pi(k)}\|_\infty^2}\right] \\ &\le 2\ exp\left[\frac{N^{2/q}(1-2/p)}{4N^{(1-2/p)}}\right] \\ &= 2\ exp\left[\frac{-(2-p)N}{4p}\right] . \end{aligned}$$

If $t \ge q+1$, set

$$N = \left[(\frac{t}{q+1})^q\right]$$

so that

$$1 \le N \le (\frac{t}{q+1})^q \le 2N .$$

Then

$$\begin{aligned} P(|f-Ef| \ge t) = P\left(|\sum_{k=1}^{n} d_k| \ge t\right] &\le P\left(|\sum_{k=1}^{n} d_k| \ge (q+1)N^{1/q}\right) \\ \le 2\ exp\left[\frac{-(2-p)N}{4p}\right] &\le 2\ exp\left[\frac{-(2-p)t^q}{8p(q+1)^q}\right] . \end{aligned}$$

If $t \le q+1$, then

$$2\ exp\left[\frac{-(2-p)t^q}{8p(q+1)^q}\right] \ge 2\ exp\left[\frac{-(2-p)}{8p}\right] \ge 2\ e^{-1/8} > 1 ,$$

and we get the desired inequality trivially in this case.

□

8.5. Consider now the probability space $\Omega = \{-1,1\}^{m\cdot n} \times \Pi_n^m$, i.e., the space of all couples $(\bar{\varepsilon},\bar{\pi})$ where $\bar{\varepsilon}$ is a matrix $\{\varepsilon_{i,j}\}_{i=1,j=1}^{m\ \ n}$ with ± 1 entries, and $\bar{\pi}$ is a vector $(\pi_1,\dots,\pi_m)$ of permutations of $\{1,\dots,n\}$, with the normalized counting measure $P(\bar{\varepsilon},\bar{\pi}) = \frac{1}{2^{m\cdot n}\cdot(n!)^m}$ for all $(\bar{\varepsilon},\bar{\pi}) \in \Omega$.

Fix $1 < p < 2$, $\varepsilon > 0$ and let $m \le \alpha n$ (α from Lemma 6.2). For each $(\bar{\varepsilon},\bar{\pi}) \in \Omega$ we define a sequence $x_1(\bar{\varepsilon},\bar{\pi}),\dots,x_m(\bar{\varepsilon},\bar{\pi})$ of vectors in ℓ_1^n by

$$x_i(\bar{\varepsilon},\bar{\pi}) = \frac{1}{n}\sum_{j=1}^{n} \varepsilon_{i,j} a_j e_{\pi_i(j)} , \quad i = 1,\dots,m$$

(a_j were defined above, $(e_j)_{j=1}^n$ is the unit vector basis in ℓ_1^n). Our purpose is to find $(\bar{\varepsilon},\bar{\pi}) \in \Omega$ such that

$$\left\|\sum_{i=1}^{m} b_i x_i(\bar{\varepsilon},\bar{\pi})\right\|_{\ell_1^n} \approx \left(\sum_{i=1}^{m} |b_i|^p\right)^{1/p}$$

for all scalars $b_1,\dots,b_m$, with m as large as possible. We shall do it in two steps. First we show (Lemma 8.6 below) that for every fixed $b_1,\dots,b_m$ this is true on the average over $(\bar{\varepsilon},\bar{\pi})\in\Omega$, then we show (Proposition 8.7) that for each $b_1,\dots,b_m$

$$\left\|\sum_{i=1}^m b_i x_i(\bar{\varepsilon},\bar{\pi})\right\|$$

is very close to its average on most of Ω. Now we use the new distributional inequality for martingales (Lemma 8.4). The estimates are such that they allow us to combine the estimates for different $b_1,\dots,b_m$ and still get good estimates for the deviation of $\|\Sigma_{i=1}^m b_i x_i(\bar{\varepsilon},\bar{\pi})\|$ from its mean, holding simultaneously for a lot of different $(b_1,\dots,b_m)$.

We choose the $(b_1,\dots,b_m)$ to form a δ-net in the unit sphere of ℓ_p^m and conclude by approximating a general vector in the sphere of ℓ_p^m by a member of this δ-net.

8.6. We begin with an estimate of the mean.

LEMMA: *Let* $\Sigma_{i=1}^m |b_i|^p = 1$. *Then*

$$1-\varepsilon \le E\left\|\sum_{j=1}^m b_j x_j(\bar{\varepsilon},\bar{\pi})\right\|_{\ell_1^n} \le 1+\varepsilon\ .$$

PROOF:

$$\begin{aligned} E\left\|\sum_{i=1}^m b_i x_i(\bar{\varepsilon},\bar{\pi})\right\| &= \frac{1}{n}E\left\|\sum_{j=1}^n\sum_{j=1}^m b_i\varepsilon_{i,j}a_j e_{\pi_i(j)}\right\| \\ &= \frac{1}{n}E\left\|\sum_{k=1}^n\sum_{i=1}^m b_i\varepsilon_{i,\pi_i^{-1}(k)}a_{\pi_i^{-1}(k)}e_k\right\| \\ &= \frac{1}{n}\sum_{k=1}^n E\left|\sum_{i=1}^m b_i\varepsilon_{i,\pi_i^{-1}(k)}a_{\pi_i^{-1}(k)}\right|\ . \end{aligned}$$

Fix $1\le k\le n$ and consider the random variables $u_1,\dots,u_m$ on Ω defined by

$$u_i(\bar{\varepsilon},\bar{\pi}) = \varepsilon_{i,\pi_i^{-1}(k)}\cdot a_{\pi_i^{-1}(k)}\ ,\qquad i=1,\dots,m\ .$$

These variables are symmetric and independent (check) and have the same distribution as $y_1,\dots,y_m$ in Lemma 8.2. So, by Lemma 8.2, for each $k=1,\dots,n$,

$$1-\varepsilon \le E\left|\sum_{i=1}^m b_i\varepsilon_{i,\pi_i^{-1}(k)}a_{\pi_i^{-1}(k)}\right| \le 1+\varepsilon$$

and the same estimates hold when taking the average over k.

□

8.7. We are now ready for the main part of the proof. For later use we state the next proposition for a general norm replacing ℓ_1^n.

PROPOSITION: *Let $1 < p < 2$, m,n positive integers, $\|\cdot\|$ a norm on $\mathbb{R}^n$, $a_1 \geq \ldots \geq a_n \geq 0$ and $x_i(\bar{\varepsilon},\bar{\pi})$ as before (i.e., $a_j = g^*(j/n)$ for g an L_1 normalized p-stable variable and $x_i(\bar{\varepsilon},\bar{\pi}) = \frac{1}{n}\Sigma_{j=1}^n \varepsilon_{i,j} a_j e_{\pi_i(j)}$) and let $\bar{b} = (b_1,\ldots,b_m)$ satisfy $\Sigma_{i=1}^m |b_i|^p = 1$. Let, for $(\bar{\varepsilon},\bar{\pi}) \in \Omega$,*

$$f(\bar{\varepsilon},\bar{\pi}) = f_{\bar{b}}(\bar{\varepsilon},\bar{\pi}) = \left\| \sum_{i=1}^m b_i x_i \right\| .$$

Then, for all $c > 0$

$$P(|f - Ef| > c) \leq 2\, exp(-\varsigma_p \cdot c^q \cdot n \cdot \max_{1\leq k\leq n} \|e_k\|^{-q})$$

where $\varsigma_p > 0$ depends only on p.

PROOF: Introduce the lexicographic order on $0 \cup \{(i,j)\}_{i=1,j=1}^{m\ \ n}$, i.e.,

$$0 < (1,1) < (1,2) < \ldots < (1,n) < (2,1) < \ldots < (2,n) < (3,1) < \ldots .$$

Let $\mathcal{F}_0 = \{\phi,\Omega\}$ and for $1 \leq i \leq m$, $1 \leq j \leq n$ define $\mathcal{F}_{(i,j)}$ to be the σ-algebra whose atoms are the sets on which $\varepsilon_{\ell,k}$ and $\pi_\ell(k)$ are constants for all $(\ell,k) \leq (i,j)$. Then $\{\mathcal{F}_t :\ t \in 0 \cup \{(i,j)\}_{i=1,j=1}^{m\ \ n}\}$ is an increasing sequence of σ-algebras the first of which is the trivial one and the last is the collection of all subsets of Ω. For $1 \leq i \leq m$, $1 \leq j \leq n$ let $(i,j)'$ be the immediate predecessor of (i,j) and define

$$d_{i,j} = E(f|\mathcal{F}_{(i,j)}) - E(f|\mathcal{F}_{(i,j)'}) .$$

In order to prove the proposition it is enough, by Lemma 8.4, to show that for all $1 \leq i \leq m$, $1 \leq j \leq n$,

$$\|\{\|d_{i,j}\|_\infty\}_{i=1,j=1}^{m\ \ n}\|_{p,\infty} \leq K_p \cdot n^{-1/q} \cdot \max_{1\leq k\leq n} \|e_k\|$$

with K_p depending only on p.

Fix $1 \leq i \leq m$, $1 \leq j \leq n$ and an atom A of $\mathcal{F}_{(i,j)'}$. The value of $E(f|\mathcal{F}_{(i,j)'})$ on A is the averaged value of $E(f|\mathcal{F}_{(i,j)})_{|B}$ where B ranges over all atoms of $\mathcal{F}_{(i,j)}$ contained in A, and, for each such B, $E(f|\mathcal{F}_{(i,j)})_{|B}$ is the averaged value of f on B. Thus, for each such B

$$|d_{i,j|B}|_\infty \leq \sup_C \left| \operatorname*{Av}_{w\in B} f(w) - \operatorname*{Av}_{w\in C} f(w) \right|$$

where the sup is taken over all atoms C of $\mathcal{F}_{(i,j)}$ contained in A.

Fix C, B atoms of $\mathcal{F}_{(i,j)}$ contained in A. Then the values of $\varepsilon_{u,v}$ and $\pi_u(v)$ are specified and equal on B and C (and A)for all $(u,v) < (i,j)$. $\varepsilon_{i,j}$ and $\pi_i(j)$ are also specified on B and on C but may be different, say,

$$\varepsilon_{i,j} = \varepsilon_B\ , \quad \pi_i(j) = s \ \text{ on } \ B$$

while

$$\varepsilon_{i,j} = \varepsilon_C\ , \quad \pi_i(j) = t \ \text{ on } \ C$$

Define a one to one and onto map $(\bar{\varepsilon}, \bar{\pi}) \to (\bar{\varepsilon}^*, \bar{\pi}^*)$ from B to C by

$$\varepsilon^*_{u,v} = \begin{cases} \varepsilon_{u,v} & \text{if} \quad (u,v) \neq (i,j) \\ \varepsilon_C & \text{if} \quad (u,v) = (i,j) \end{cases}$$

$$\pi^*_u = \begin{cases} \pi_u & \text{if} \quad u \neq i \\ \rho \circ \pi_u & \text{if} \quad u = i \end{cases}$$

where ρ replaces s with t and leaves the rest in place. Then

$$\begin{aligned} |f(\bar{\varepsilon}, \bar{\pi}) - f(\bar{\varepsilon}^*, \bar{\pi}^*)| &= \frac{1}{n} \left\| \sum_{u=1}^{m} \sum_{v=1}^{n} b_u \varepsilon_{u,v} a_v e_{\pi_u(v)} - b_u \varepsilon^*_{u,v} a_v e_{\pi^*_u(v)} \right\| \\ &= \frac{1}{n} \left\| \sum_{v=1}^{n} b_i a_v [\varepsilon_{i,v} e_{\pi_i(v)} - \varepsilon^*_{i,v} e_{\pi^*_i(v)}] \right\| \end{aligned}$$

the two expressions in the parenthesis are equal except possibly when $v = j$ and when $v = \pi_i^{-1}(t)$ so, by the triangle inequality,

$$|f(\bar{\varepsilon}, \bar{\pi}) - f(\bar{\varepsilon}^*, \bar{\pi}^*)| \leq \frac{1}{n} |b_i|\, |a_j| [\|e_t\| + \|e_s\|] + \frac{1}{n} |b_i|\, |a_{\pi_i^{-1}(t)}| [\|e_s\| + \|e_t\|] .$$

Notice that $\pi_i^{-1}(t) \geq j$ so that $|a_{\pi_i^{-1}(t)}| \leq |a_j|$ and

$$|f(\bar{\varepsilon}, \bar{\pi}) - f(\bar{\varepsilon}^*, \bar{\pi}^*)| \leq \frac{4}{n} |b_i|\, |a_j| \max_{1 \leq k \leq n} \|e_k\| .$$

It follows that,

$$\left| \operatorname*{Av}_{w \in B} f(w) - \operatorname*{Av}_{w \in C} f(w) \right| \leq \frac{4}{n} |b_i|\, |a_j| \max_{1 \leq k \leq n} \|e_k\|$$

and

$$\|d_{i,j}\|_\infty \leq \frac{4}{n} |b_i|\, |a_j| \max_{1 \leq k \leq n} \|e_k\| .$$

By the estimate on the tail of the distribution of a p-stable (see the proof of Lemma 8.2) we get

$$|a_j| = g^*\left(\frac{j}{n}\right) \leq C^{-1/p} \frac{n^{1/p}}{j^{1/p}}$$

and

$$\|\{a_j\}_{j=1}^n\|_{p,\infty} \leq (n/C)^{1/p} .$$

By the upper p-estimate property of $\|\cdot\|_{p,\infty}$

$$\begin{aligned} \|\{b_i a_j\}_{i=1,j=1}^{n\ \ m}\|_{p,\infty} &\leq \left(\sum_{i=1}^{n} \|\{b_i a_j\}_{j=1}^m\|_{p,\infty}^p \right)^{1/p} \\ &\leq C^{-1/p} \left(\sum_{i=1}^{n} |b_i|^p \cdot n \right)^{1/p} = (n/C)^{1/p} \end{aligned}$$

and we get

$$\left\|\{\|d_{i,j}\|_\infty\}_{i=1,j=1}^{n\ \ m}\right\|_{p,\infty} \le 4C^{-1/p}n^{(1/p)-1} \max_{1\le k\le n} \|e_k\| = 4C^{-1/p}n^{-1/q} \max_{1\le k\le n} \|e_k\| .$$

□

8.8. THEOREM: *Let $1 < p < 2$, $0 < \varepsilon$. Then there exists an $\beta = \beta(\varepsilon, p)$ such that if $m \le \beta n$ then there exists a subspace $X \subseteq \ell_1^n$, dim $X = m$, and $d(X, \ell_p^m) \le 1 + \varepsilon$.*

PROOF: With notations as above we get, from Proposition 8.7 and Lemma 8.6 (here $\{e_k\}_{k=1}^n$ are the unit vector basis of ℓ_1^n),

$$P(1 - \varepsilon - c \le f_{\bar b} \le 1 + \varepsilon + c) \le 1 - 2\, exp(-\varsigma_p \cdot c^q \cdot n)$$

for all $\bar b = (b_1, \ldots, b_m)$ in $S(\ell_p^m) = \{\bar b = (b_1, \ldots, b_m);\ \sum_{i=1}^m |b_i|^p = 1\}$.

The rest of the proof is standard:

If $N \subseteq S(\ell_p^m)$,

$$P\left(1 - \varepsilon - c \le \left\|\sum_{i=1}^m b_i x_i(\bar\varepsilon, \bar\pi)\right\| \le 1 + \varepsilon + c \text{ for all } \bar b \in N\right) \ge 1 - 2|N| exp(-\varsigma_p \cdot c^q \cdot n) .$$

The right hand side is positive as long as $|N| < \frac{1}{2} exp(\varsigma_p \cdot c^q \cdot n)$ so that N can be chosen to be a δ-net in the sphere of ℓ_p^m if $m \le \beta(\delta, p, c) \cdot n$. If $c = \frac{\varepsilon}{2}$ and δ is chosen small with respect to ε then

$$1 - \frac{\varepsilon}{2} \le \left\|\sum_{i=1}^m b_i x_i(\bar\varepsilon, \bar\pi)\right\| \le 1 + \frac{3}{2}\varepsilon$$

for all $\bar b$ in N implies a similar inequality for all $\bar b$ in the sphere of ℓ_p^m (see for example the proof of Lemma 4.1).

□

9. TYPE AND COTYPE OF NORMED SPACES, AND SOME SIMPLE RELATIONS WITH GEOMETRICAL PROPERTIES

Let X be a normed space, $x_i \in X$, $i = 1, 2, \dots$. In this section we will study the averages $(\operatorname{Ave}_{\varepsilon_i=\pm 1} \|\Sigma_{i=1}^n \varepsilon_i x_i\|^2)^{1/2}$ and see that the order of magnitude of these averages gives a lot of information about some geometrical properties of X.

We shall sometimes use a different way to write these averages

$$\operatorname*{Ave}_{\varepsilon_i=\pm 1} \left\| \sum_{i=1}^n \varepsilon_i x_i \right\|^2 = \int_0^1 \left\| \sum_{i=1}^n r_i(t) x_i \right\|^2 dt$$

where r_i are the Rademacher functions (see 5.5).

9.1. Given a normed space X, a natural number n, and $1 \le p \le 2$ (resp. $2 \le q < \infty$), let $T_p(X, n)$ (resp. $C_q(X, n)$) be the smallest T (resp. C) such that

$$\left(\int_0^1 \left\| \sum_{i=1}^n r_i(t) x_i \right\|^2 dt \right)^{1/2} \le T \left(\sum_{i=1}^n \|x_i\|^p \right)^{1/p}$$

$$\left(resp. \quad \left(\sum_{i=1}^n \|x_i\|^q \right)^{1/q} \le C \left(\int_0^1 \left\| \sum_{i=1}^n r_i(t) x_i \right\|^2 \right)^{1/2} \right)$$

for all $x_1, \dots, x_n \in X$.

Let $T_p(X) = \sup_n T_p(X, n)$, $C_q(X) = \sup_n C_q(X, n)$. If $T_p(X) < \infty$ (resp. $C_q(X) < \infty$) we say that X has *type p* and/or that the *type p constant of* X is $T_p(X)$ (resp. X has *cotype q constant* $C_q(X)$). When there is no confusion about the X we are dealing with we shall omit the symbol X. We have seen before (5.5) that for a Hilbert space H, $T_2(H) = C_2(H) = 1$.

9.2. Kahane's inequality. For any $1 \le p < \infty$ there exists a constant K_p such that

$$\int_0^1 \left\| \sum_{i=1}^n r_i(t) x_i \right\| dt \le \left(\int_0^1 \left\| \sum_{i=1}^n r_i(t) x_i \right\|^p dt \right)^{1/p} \le K_p \int_0^1 \left\| \sum_{i=1}^n r_i(t) x_i \right\| dt .$$

This inequality seems to be closely related to the fact that $E_2^n = \{-1, 1\}^n$ is a Levy family. However, we are not aware of a proof along these lines. A proof of inequality 9.2 is given in Appendix III.

9.3. EXAMPLES: 1. L_p, $1 \le p \le 2$, has type p and cotype 2.

PROOF. We use the notation $\approx$ when the expressions on the two sides of sign are equivalent up to a constant depending on p alone (in this proof – the constants from Khinchine's inequality or Kahane's inequality).

For $x_i \in L_p(\Omega,\mu)$, $i = 1,\ldots,n$,

$$\left(\int_0^1 \left\|\sum_{i=1}^n r_i(t)x_i\right\|^2 dt\right)^{p/2} \approx \int_0^1 \left\|\sum_{i=1}^n r_i(t)x_i\right\|^p dt$$

$$= \int_0^1 \int_\Omega \left|\sum_{i=1}^n r_i(t)x_i(\omega)\right|^p = \int_\Omega \int_0^1 \left|\sum_{i=1}^n r_i(t)x_i(\omega)\right|^p dt\, d\mu(\omega)$$

$$\approx \int_\Omega \left(\sum_{i=1}^n x_i^2(\omega)\right)^{p/2} d\mu .$$

(For further use, note that up to now the equivalence holds for all $1 \le p < \infty$.)

Now, since $\|\cdot\|_{\ell_2} \le \|\cdot\|_{\ell_p}$

$$\int_\Omega \left(\sum_{i=1}^n x_i^2(\omega)\right)^{p/2} d\mu \le \int_\Omega \sum_{i=1}^n |x_i(\omega)|^p d\mu = \sum_{i=1}^n \|x_i\|^p$$

which proves that L_p has type p. To prove cotype 2, use the triangle inequality (for integrals rather than sums) for the norm in $\ell_{2/p}$

$$\int_\Omega \left(\sum_{i=1}^n x_i^2(\omega)\right)^{p/2} d\mu = \int_\Omega \left(\sum_{i=1}^n |x_i(\omega)|^{p\cdot(2/p)}\right)^{p/2} d\mu$$

$$\ge \left(\sum_{i=1}^n \left(\int_\Omega |x_i(\omega)|^p d\mu\right)^{2/p}\right)^{p/2} = \left(\sum_{i=1}^n \|x_i\|^2\right)^{p/2}$$

□

2. L_q, $2 \le q < \infty$ has type 2 and cotype q. We leave the proof for the reader.

9.4. One may consider averages with respect to random variables other than the Rademacher functions. Of particular interest are the gaussian averages. We recall that a normalized in L_2 gaussian variable is a random variable $g(\omega)$ whose distribution is given by

$$P(g(\omega) \le t) = \frac{1}{\sqrt{2\pi}} \int_{-\infty}^t e^{-s^2/2} ds .$$

Let $\{g_i\}_{i=1}^\infty$ be a sequence of independent gaussian random variables normalized in L_2. For $1 \le p \le 2 \le q < \infty$ and $n \in N$ we define the *gaussian type* p (resp. *cotype* q) constants $\alpha_p(X,n)$ (resp. $\beta_q(X,n)$) of X as the smallest T (resp. C) such that

$$\left(E\left\|\sum_{i=1}^n g_i(\omega)x_i\right\|^2\right)^{1/2} \le T\left(\sum_{i=1}^n \|x_i\|^p\right)^{1/p}$$

$$\left(\textit{resp.}\quad \left(\sum_{i=1}^n \|x_i\|^q\right)^{1/q} \le C\left(E\left\|\sum_{i=1}^n g_i(\omega)x_i\right\|^2\right)^{1/2}\right)$$

$$\left(resp. \quad \left(\sum_{i=1}^{n} \|x_i\|^q \right)^{1/q} \leq C \left(E \| \sum_{i=1}^{n} g_i(\omega) x_i \|^2 \right)^{1/2} \right)$$

for all $x_1, \ldots, x_n \in X$.

We want to compare $\alpha_p(X,n)$ and $\beta_q(X,n)$ with $T_p(X,n)$ and $C_q(X,n)$. First note that the distribution of $g_1(\omega), \ldots, g_n(\omega)$ is the same as that of $r_i(t)|g_i(\omega)|, \ldots, r_n(t)|g_n(\omega)|$, so, by convexity,

$$\begin{aligned}
\left(E \| \sum_{i=1}^{n} g_i(\omega) x_i \|^2 \right)^{1/2} &= \left(E \| \sum_{i=1}^{n} r_i(t) |g_i(\omega)| x_i \|^2 \right)^{1/2} \\
&\geq \left(E_t \| \sum_{i=1}^{n} r_i(t) E_\omega |g_i(\omega)| x_i \|^2 \right)^{1/2} \\
&= \sqrt{2/\pi} \cdot \left(E \| \sum_{i=1}^{n} r_i(t) x_i \|^2 \right)^{1/2}
\end{aligned}$$

and we get

9.4.1. $$T_p(X,n) \leq \sqrt{\pi/2} \cdot \alpha_p(X,n)$$

and

9.4.2. $$\beta_q(X,n) \leq \sqrt{\pi/2} \cdot C_q(X,n).$$

Next, notice that

$$\left(E \| \sum_{i=1}^{n} g_i x_i \|^2 \right)^{1/2} = \left(E \| \sum_{i=1}^{n} r_i(t) g_i(\omega) x_i \|^2 \right)^{1/2}$$

$$\leq T_2(X,n) \left(E \sum_{i=1}^{n} (|g_i(\omega)| \, \|x_i\|)^2 \right)^{1/2} = T_2(X,n) \left(\sum_{i=1}^{n} \|x_i\|^2 \right)^{1/2}$$

so that

9.4.3. $$\alpha_2(X,n) \leq T_2(X,n).$$

Using Kahane's inequality one gets similarly that, for all $1 \leq p \leq 2$,

9.4.4. $$\alpha_p(X,n) \leq K_p \cdot T_p(X,n).$$

Also, a similar computation gives,

9.4.5. $$\beta_2(X,n) \leq C_2(X,n)$$

and

9.4.6. $$\beta_q(X,n) \leq K_q \cdot C_q(X,n)$$

(which is better than $\beta_q(X,n) \leq \sqrt{\pi/2} \cdot C_q(X,n)$ for $q = 2$ and other values of q close to 2).

The remaining inequality, namely $C_q(X,n) \leq K_q \cdot \beta_q(X,n)$ does not hold in general. We discuss this matter in Appendix II.

9.5. The next simple computation shows the relation between the gaussian average and another average we have previously used: integration with respect to μ— the Haar measure on S^{n-1}.

$$E\|\sum_{i=1}^n g_i(\omega)x_i\|^2 = \frac{1}{(2\pi)^{n/2}} \int_{\mathbb{R}^n} \|\sum_{i=1}^n t_i x_i\|^2 exp\left(-\sum_{i=1}^n \frac{t_i^2}{2}\right) dt_1 \dots dt_n$$
$$= C_n \cdot \int_{S^{n-1}} \|\sum_{i=1}^n t_i x_i\|^2 d\mu \int_0^\infty R^{n+1} exp(-R^2/2) dR$$
$$= C_n' \cdot \int_{S^{n-1}} \|\sum_{i=1}^n t_i x_i\|^2 d\mu \ .$$

C_n and C_n' depend only on n so one can compute them using $(x_i)_{i=1}^n$ an orthonormal system in a Hilbert space and get

$$n = C_n' \int_{S^{n-1}} (\Sigma t_i^2) d\mu = C_n'$$

i.e.

9.5.1.
$$E\|\sum_{i=1}^n g_i(\omega)x_i\|^2 = n \int_{S^{n-1}} \|\sum_{i=1}^n t_i x_i\|^2 d\mu \ .$$

We now bring two theorems relating the notions of type and cotype to geometry.

9.6. THEOREM: *Let X be a n-dimensional normed space with cotype q constant C_q for some $q < \infty$. Then $k(X) \geq c \cdot C_q^{-2} \cdot n^{2/q}$ for some absolute constant c. ($k(X)$ was defined in 4.3).*

PROOF: Let $|\cdot|$ be the euclidean norm on X given by the ellipsoid of maximal volume contained in $B(X)$. Then, by Theorem 3.4, there exists an orthonormal basis $\{e_i\}_{i=1}^n$ with $\|e_i\| \geq 1/4$ for $1 \leq i \leq n/2$. Therefore (by Theorems 5.1 and 3.3)

$$M_{\|\cdot\|} \approx \left(\int_{S^{n-1}} \|\Sigma a_i e_i\|^2 d\mu(\bar{a})\right)^{1/2}$$
$$= \left(\int_{S^{n-1}} \int_0^1 \|\sum_{i=1}^n a_i r_i(t) e_i\|^2 dt \ d\mu(\bar{a})\right)^{1/2}$$
$$\geq (2\cdot C_q)^{-1} \left(\int_{S^{n-1}} (\sum_{i=1}^{[n/2]} |a_i|^q)^{2/q} d\mu(\bar{a})\right)^{1/2}$$
$$\geq (2\cdot C_q)^{-1} (n/2)^{\frac{1}{q}-\frac{1}{2}} \left(\int_{S^{n-1}} \sum_{i=1}^{[n/2]} |a_i|^2 d\mu(\bar{a})\right)^{1/2}$$
$$= (2\cdot C_q)^{-1} \left(\frac{[n/2]}{n}\right)^{1/2} \cdot n^{\frac{1}{q}-\frac{1}{2}}$$

and by Theorem 4.2,

$$k(X) \geq c_1 \cdot n \cdot M_r^2 \geq c \cdot C_q^{-2} \cdot n^{2/q} .$$

□

REMARK: The proof actually shows that the factor C_q can be replaced by $C_q(X,n)$.

9.7. THEOREM: *Let X be a normed space whose gaussian type 2 constant on n vectors is $\alpha(n) = \alpha(X,n)$. Then*

$$k(X) \geq c \cdot \alpha(n)^2$$

for some absolute constant c.

PROOF: We actually prove a somewhat stronger theorem: If

$$E\|\sum_{j=1}^{n} g_j x_j\| \geq \alpha \cdot \left(\sum_{j=1}^{n} \|x_j\|^2\right)^{1/2}$$

for some $\{x_j\}_{j=1}^n \subseteq X$, not all of which are zero, then $k(span\{x_i\}_{i=1}^n) \geq c \cdot \alpha^2$.

Fix $m, k \in \mathbb{N}$ and consider the probability space $\{-1,1\}^{k \cdot m \cdot n}$ with the uniform measure. For $\bar{\varepsilon} \in \{-1,1\}^{k \cdot m \cdot n}$ define

$$u_i(\bar{\varepsilon}) = \frac{1}{\sqrt{k}} \sum_{j=1}^{n} \sum_{\ell=1}^{k} \varepsilon_{i,j,\ell} x_j$$

and, given independent, normalized gaussian variables $\{g_{i,j}\}_{i=1,j=1}^{m}{}_{}^{n}$, define

$$v_i(\omega) = \sum_{j=1}^{k} g_{i,j}(\omega) x_j .$$

Note that the central limit theorem implies that the distribution of $(u_1, \ldots, u_m)$ tends, as $k \to \infty$, to that of $(v_1, \ldots, v_m)$.

Let $\bar{a} = (a_1, \ldots, a_m)$ be such that $\Sigma_{i=1}^m a_i^2 = 1$ and consider the function

$$f(\bar{\varepsilon}) = f_{\bar{a}}(\bar{\varepsilon}) = \|\sum_{i=1}^{m} a_i u_i(\bar{\varepsilon})\| .$$

We form the obvious sequence of $k \cdot m \cdot n$ σ fields $\mathcal{F}_{i,j,\ell}$ (i.e. an atom of $\mathcal{F}_{i,j,\ell}$ is a set where all of $\varepsilon_{u,v,w}$ are specified for all $(u,v,w) \leq (i,j,\ell)$ in some pre-specified, say the lexicographic, order) and we get a martingale

$$f_{i,j,\ell} = E(f|\mathcal{F}_{i,j,\ell})$$

with martingale differences $(d_{i,j,\ell})_{i=1,j=1,\ell=1}^{m\ \ n\ \ k}$.

We leave it to the reader to check that

$$\|d_{i,j,\ell}\|_\infty \leq 2|a_i| \frac{1}{\sqrt{k}} \|x_j\| .$$

By Lemma 7.4 we get

$$P\left(\|\sum_{i=1}^{m} a_i u_i(\overline{\varepsilon})\| - E\|\sum_{i=1}^{m} a_i u_i(\cdot)\| > c\right) \le 2\ exp\left(\frac{-c^2}{16\Sigma_{j=1}^{n}\|x_j\|^2}\right) .$$

Let $k \to \infty$ to obtain

$$P\left(\|\sum_{i=1}^{m} a_i v_i(\omega)\| - E\|\sum_{i=1}^{m} a_i v_i(\cdot)\| > c\right) \le 2\ exp\left(\frac{-c^2}{16\Sigma_{j=1}^{n}\|x_j\|^2}\right) .$$

Note that since $\Sigma_{i=1}^{m} a_i^2 = 1$, $(\Sigma_{i=1}^{m} a_i g_{i,1}, \ldots, \Sigma_{i=1}^{m} a_i g_{i,n})$ is again a sequence of independent gaussian variables normalized in L_2 so that

$$E\|\sum_{i=1}^{m} a_i v_i(\cdot)\| = E\|\sum_{j=1}^{n} g_j x_j\| .$$

Choosing $c = \varepsilon \cdot E\|\Sigma_{j=1}^{n} g_i x_i\|$ we get

$$P\left((1-\varepsilon)E\|\sum_{j=1}^{n} g_j x_j\| \le \|\sum_{i=1}^{m} a_i v_i(\omega)\| \le (1+\varepsilon)E\|\sum_{j=1}^{n} g_j x_j\|\right) \ge 1 - 2\ exp\left(\frac{-\varepsilon^2\alpha^2}{16}\right) .$$

The rest of the proof is standard (see the proof of Lemma 4.1).

□

9.8. Given a measure space $(\Omega, \mathcal{F}, \mu)$ and a Banach space X, we define the space $L_2(X) = L_2(X, \Omega)$ as the space of all measurable functions from $(\Omega, \mathcal{F})$ to X with

$$\|f\| = \left(\int_\Omega \|f(\omega)\|_X^2 d\mu\right)^{1/2} < \infty .$$

In the cases we shall be interested in, the space Ω will be finite and no problem of measurability will arise. Actually, the only measure space relevant to us is the space $\{-1,1\}^n$ with μ – the normalized counting measure.

For $f \in L_2(X)$, $g \in L_2(X^*)$ we define $< g, f >= \int g(\omega)(f(\omega))d\mu$. This defines a duality relation. Dimension consideration actually shows that $L_2(X)^* = L_2(X^*)$ with equality of norm $\|\cdot\|_{L_2(X)}^* = \|\cdot\|_{L_2(X^*)}$, when X is finite dimensional.

9.9. Given any orthonormal basis $\{w_\alpha\}_{\alpha \in A}$ for $L_2(\{-1,1\}^n, \mu)$ one can expand any $f \in L_2(X)$ as $f(t) = \Sigma_{\alpha \in A} w_\alpha(t) x_\alpha$ with some $\{x_\alpha\}_{\alpha \in A} \subseteq X$. Indeed $x_\alpha = \int f(t) w_\alpha(t) d\mu(t)$ (note:this integral is actually a finite sum). We will be interested mostly in one particular basis: the Walsh system. Let $(r_i(t))_{i=1}^n$ be the Rademacher functions on $\{-1,1\}^n$, i.e., $r_i(t) = r_i(t_1, \ldots, t_n) = t_i$. For $A \subseteq \{1, \ldots, n\}$, $A \neq \emptyset$, let

$$w_A(t) = \prod_{i \in A} r_i(t)$$

and let

$$w_{\emptyset}(t) \equiv 1 \ .$$

It is easily checked that $\{w_A\}_{A\subseteq\{1,\dots,n\}}$ is an orthonormal system of 2^n functions in a space of dimension 2^n so it must be complete. Note that $r_i = w_{\{i\}}$.

There are several other ways to introduce the Walsh system: Let $w_0 = 1$ and inductively define the $2^n \times 2^n$ matrix W_n (called the Walsh matrix) by

$$W_n = \begin{pmatrix} W_{n-1} & W_{n-1} \\ W_{n-1} & -W_{n-1} \end{pmatrix}$$

(for example

$$W_1 = \begin{pmatrix} 1 & 1 \\ 1 & -1 \end{pmatrix}, \quad W_2 = \begin{pmatrix} 1 & 1 & 1 & 1 \\ 1 & -1 & 1 & -1 \\ 1 & 1 & -1 & -1 \\ 1 & -1 & -1 & 1 \end{pmatrix}, \dots \ ).$$

Then one can identify the points $\{-1,1\}^n$ with $(1,\dots,2^n)$ in such a manner that $(W_A(t))_{A\subseteq\{1,\dots,n\}}$ are identified with the 2^n rows of W_n.

Alternatively, $\{w_A\}_{A\subseteq\{1,\dots,n\}}$ is the group of characters of the multiplicative group $\{-1,1\}^n$.

9.10. We will be interested in one special subspace of $L_2(X,\{-1,1\}^m)$ – the one spanned by the first n Rademacher functions:

$$Rad_n X = \left\{ \sum_{i=1}^{n} r_i(t) x_i \ ; \quad x_i \in X \ , \quad i = 1,\dots n \right\} \ .$$

There are natural projections onto these subspaces; for $f = \Sigma_{A\subseteq\{1,\dots,m\}} w_A \cdot x_A$ we define

$$Rad_n f = \sum_{i=1}^{n} r_i \cdot x_{\{i\}} \ .$$

The norms of these projections play a central role in the theory of type and cotype.

LEMMA: *Let X be a normed space, n an integer and let $1 < p \le 2 \le q < \infty$ with $\frac{1}{p} + \frac{1}{q} = 1$. Then*

$$C_q(X,n) \le T_p(X^*,n) \le \|Rad_n\| C_q(X,n) \ .$$

PROOF: For any $g(t) = \Sigma_{i=1}^n r_i(t) x_i^*$, $x_i^* \in X^*$,

$$\|g\|_{L_2} = \left(\int \| \sum_{i=1}^{n} r_i(t) x_i^* \|^2 d\mu \right)^{1/2} \le T_p(X^*,n) \left(\sum_{i=1}^{n} \|x_i^*\|^p \right)^{1/p} \ .$$

Given $x_i \in X,\ i=1,\dots,n$, let $x_i^* \in X^*,\ i=1,\dots,n$, be such that $x_i^*(x_i) = \|x_i^*\| \cdot \|x_i\|$ and $\|x_i^*\|^p = \|x_i\|^q$. Then

$$\left(\sum_{i=1}^n \|x_i\|^q\right)^{1/q} \left(\sum_{i=1}^n \|x_i^*\|^p\right)^{1/p} = \sum_{i=1}^n \|x_i^*\| \cdot \|x_i\| = \sum_{i=1}^n x_i^*(x_i)$$

$$=< \sum_{i=1}^n r_i(t)x_i^*, \sum_{i=1}^n r_i(t)x_i > \leq \|\sum_{i=1}^n r_i(t)x_i^*\|_{L_2(X^*)} \cdot \|\sum_{i=1}^n r_i(t)x_i\|_{L_2(X)}$$

$$\leq T_p(X^*,n)\left(\sum_{i=1}^n \|x_i^*\|^p\right)^{1/p} \cdot \|\sum_{i=1}^n r_i(t)x_i\|_{L_2(X)} .$$

Therefore,

$$\left(\sum_{i=1}^n \|x_i\|^q\right)^{1/q} \leq T_p(X^*,n)\left(\int \|\sum_{i=1}^n r_i(t)x_i\|^2 d\mu\right)^{1/2} .$$

Thus,

$$C_q(X,n) \leq T_p(X^*,n) .$$

To prove the right hand side inequality, note first that for any

$$f(t) = \sum_{i=1}^m r_i(t)x_i + \sum_{|A|\neq 1} w_A(t)x_A \in L_2(X) ,$$

$$\|f(t)\|_{L_2(X)} \geq \|Rad_n\|^{-1} \cdot \|\sum_{i=1}^n r_i(t)x_i\|_{L_2(X)}$$

$$\geq \|Rad_n\|^{-1} \cdot C_q(X,n)^{-1}\left(\sum_{i=1}^n \|x_i\|^q\right)^{1/q} .$$

Given $x_i^* \in X^*,\ i=1,\dots,n$, let

$$f(t) = \sum_{i=1}^m r_i(t)x_i + \sum_{|A|\neq 1} w_A(t)x_A \in L_2(X)$$

be such that $\|f\|_{L_2(X)} = 1$ and

$$\|\sum_{i=1}^n r_i(t)x_i^*\|_{L_2(X^*)} =< \sum_{i=1}^n r_i(t)x_i^*,\ f(t) > .$$

Then

$$\|\sum_{i=1}^n r_i(t)x_i^*\|_{L_2(X^*)} =< \sum_{i=1}^n r_i(t)x_i^*,\ f(t) >$$

$$= \sum_{i=1}^n x_i^*(x_i) \leq \left(\sum_{i=1}^n \|x_i^*\|^p\right)^{1/p} \cdot \left(\sum_{i=1}^n \|x_i\|^q\right)^{1/q}$$

$$\leq \|Rad_n\| \cdot C_q(X,n) \cdot \|f(t)\|_{L_2(X)} \cdot \left(\sum_{i=1}^n \|x_i^*\|^p\right)^{1/p}$$

$$= \|Rad_n\| \cdot C_q(X,n) \cdot \left(\sum_{i=1}^n \|x_i^*\|^p\right)^{1/p} .$$

Thus,

$$T_p(X^*,n) \leq \|Rad_n\| C_q(X,n) .$$

□

9.11. As an immediate corollary we get that, if X^* has type p, X has cotype q $(\frac{1}{p}+\frac{1}{q}=1)$ and that if $\sup_n \|Rad_n\| < \infty$ then also the converse holds: if X has cotype q then X^* has type p.

The assumption $\sup_n \|Rad_n\| < \infty$ is really needed: ℓ_1 has cotype 2 while ℓ_∞ has no type > 1.

We shall return to the subject of estimating $\|Rad_n\|$ in Chapter 14.

CHAPTER 10: ADDITIONAL APPLICATIONS OF LEVY FAMILIES IN THE THEORY OF FINITE DIMENSIONAL NORMED SPACES

We saw in the previous chapters some applications of the concentration of measure phenomenon on S^n. For example, we have proved in Chapters 4, 5 and 9 Dvoretzky's Theorem and some of its generalizations. We have also seen applications of the same phenomenon and of some martingale techniques in the families $\{E_2^n\}_{n=1}^{\infty}$ and $\{\Pi_n\}_{n=1}^{\infty}$ (see e.g., Theorem 9.7 and most of Chapter 8). In this chapter we will see a somewhat different kind of applications using some other Levy families. Some of the examples will be described only briefly leaving most of the details to the reader.

10.1. Let (X,ρ,μ) be a compact metric space with a Borel probability measure μ and let f be a continuous function on X, $f \in C(X)$. As we have seen in Chapter 5, it is useful to compare the median M_f with the average $E(f) = \int_X f d_\mu$. To estimate the difference $M_f - E(f)$ we use the same approach as in Lemma 5.1.

Let

$$\alpha_X(\varepsilon) = sup\{1 - \mu(A_\varepsilon);\ A \subseteq X,\ \mu(A) \geq \frac{1}{2}\}.$$

Take any partition $0 = \varepsilon_0 < \varepsilon_1 \ldots < \varepsilon_N = diam\ X$ and put $\delta_i = \omega_f(\varepsilon_i)$ where, as before, $\omega_f(\cdot)$ is the modulus of continuity of f. Then

$$|M_f - E(f)| \leq \sum_{i=0}^{N-1} \int_{\delta_i < |M_f - f| < \delta_{i+1}} |M_f - f| d\mu \leq$$

$$\leq \sum_{i=0}^{N-1} \delta_{i+1} \cdot \mu\{x; |M_f - f(x)| > \delta_i\} \leq 2 \sum_{i=0}^{N-1} \omega_f(\varepsilon_{i+1}) \cdot \alpha_X(\varepsilon_i)\ .$$

For example, in the typical case where $\alpha_X(\varepsilon) \leq c_1 \cdot exp(-c_2 \cdot \varepsilon^2 \cdot n)$ and $\omega_f(\varepsilon) \leq C \cdot \varepsilon$ we get, for any $\varepsilon > 1/\sqrt{n}$ (choosing $\varepsilon_m = \varepsilon \cdot m$),

$$|M_f - E(f)| \leq 2 \cdot C \cdot \varepsilon \cdot c_1 \cdot \sum_{m=0}^{\infty} (m+1) exp(-c_2 \cdot m^2 \cdot \varepsilon^2 \cdot n) \leq K \cdot C \cdot \varepsilon$$

where $K = K(c_1, c_2)$.

10.2. Let $(X, \|\cdot\|)$ be a normed space. A sequence $\{x_i\}_{i=1}^k$ in $X \setminus \{0\}$ is said to be *K-unconditional* (resp. *K-symmetric*) if for all $\{a_i\}_{i=1}^k \in \mathbb{R}^k$ and all $\varepsilon_i = \pm 1,\ i = 1, \ldots, k$,

$$\left\| \sum_{i=1}^k \varepsilon_i a_i x_i \right\| \leq K \left\| \sum_{i=1}^k a_i x_i \right\|$$

(resp. for all $\{a_i\}_{i=1}^k \in \mathbb{R}^k$, all $\varepsilon_i = \pm 1$, $i = 1,\dots,k$ and all $\pi \in \Pi_k$

$$\left\|\sum_{i=1}^k \varepsilon_i a_{\pi(i)} x_i\right\| \le K \left\|\sum_{i=1}^k a_i x_i\right\|).$$

The importance of unconditional and symmetric sequences stems from the fact that a lot of the desired structure theory, which does not hold in general normed spaces, holds in spaces spanned by such sequences (see [L.T-I,II] and references there).

10.3. In our first application in this chapter we use the normal Levy family from 6.5.2: $X = \Pi_{i=1}^m S_{(i)}^{n-1}$ where each of $S_{(i)}^{n-1}$ is the euclidean sphere S^{n-1}. The metric is $\rho(\bar{x},\bar{y}) = \left(\Sigma_{i=1}^m \rho_i(x_i,y_i)^2\right)^{1/2}$ (ρ_i the geodesic metric on $S_{(i)}^{n-1}$), for $\bar{x} = (x_1,\dots,x_m)$, $\bar{y} = (y_1,\dots,y_m)$, and the measure μ is the product of the Haar measures.

THEOREM: *Let E be a (finite or infinite dimensional) Banach space with cotype q constant C_q, for some $q < \infty$. Denote $S(E) = \{x \in E; \|x\| = 1\}$ and let*

$$\varphi : S^{n-1} \to S(E)$$

be an antipodal map (i.e. $\varphi(-x) = -\varphi(x)$) with Lipschitz constant L (i.e. $\|\varphi(x) - \varphi(y)\| \le L\cdot\rho(x,y)$). Then, for each $\eta > 0$, there exists a sequence $\{x_i\}_{i=1}^m \subseteq S^{n-1}$ such that $\{\varphi(x_i)\}_{i=1}^m$ forms a $(1+\eta)$-symmetric sequence in E and

$$m \ge c(\eta)\cdot(L\cdot C_q)^{-q/(q-1)} n^{q/2(q-1)}$$

where $c(\eta) > 0$ depends on η alone. If E has no cotype the same conclusion holds with $q = \infty$ (in that case $q/(q-1) = 1$ and $C_q = 1$).

PROOF: For $\bar{a} = (a_1,\dots,a_m) \in \mathbb{R}^m$ and $\bar{x} = (x_1,\dots,x_m) \in X$ $(= \Pi_{i=1}^m S_{(i)}^{n-1})$, let

$$f_{\bar{a}}(\bar{x}) = \left\|\sum_{i=1}^m a_i \varphi(x_i)\right\| \quad \textit{and} \quad E(f_{\bar{a}}) = \int_X f_{\bar{a}}(\bar{x})d\mu(\bar{x}).$$

The function $\|\|\bar{a}\|\| = E(f_{\bar{a}})$ on $\mathbb{R}^m$ is a norm. It is easily seen that the natural basis is 1-symmetric in this norm. We are going to show, using the concentration of measure phenomenon on X, that, for some $\bar{x}$, $f_{\bar{a}}(\bar{x})$ is close to $\|\|\bar{a}\|\|$ for all $\bar{a} \in \mathbb{R}^m$. Then $\{x_i\}_{i=1}^m$ will be the desired sequence. We devide the argument into several steps.

10.3.1. Let $M_{\bar{a}}$ be the median of $f_{\bar{a}}$ on X and $\omega_{\bar{a}}(\cdot)$ its modulus of continuity. Then, by Theorem 6.5.2,

$$\mu\{\bar{x}; |f_{\bar{a}}(\bar{x}) - M_{\bar{a}}| < \omega_{\bar{a}}(\varepsilon)\} \ge 1 - \sqrt{\pi/2}\, exp(-\varepsilon^2\cdot n/2).$$

Now,

$$\omega_{\bar{a}}(\varepsilon) \le \sup_{\Sigma\varepsilon_i^2 \le \varepsilon^2} \sum_{i=1}^m |a_i|\cdot L\cdot\varepsilon_i \le L\cdot\varepsilon\cdot\left(\sum_{i=1}^m a_i^2\right)^{1/2} \le L\cdot\varepsilon\cdot m^{\frac{1}{2}-\frac{1}{q}}\left(\sum_{i=1}^m |a_i|^q\right)^{1/q}$$

and we get

$$\mu\{\bar{x}; |f_{\bar{a}}(\bar{x}) - M_{\bar{a}}| < L\cdot\varepsilon\cdot m^{\frac{1}{2}-\frac{1}{q}}\cdot\left(\sum_{i=1}^{m}|a_i|^q\right)^{1/q}\} \geq 1-\sqrt{\pi/2}\ exp(-\varepsilon^2\cdot n/2).$$

10.3.2. Now we replace $M_{\bar{a}}$ by $E(f_{\bar{a}}) = |||\bar{a}|||$. Since, by 10.1,

$$|M_{\bar{a}} - |||\bar{a}|||| \leq K\cdot L\cdot\varepsilon\cdot m^{\frac{1}{2}-\frac{1}{q}}\cdot\left(\sum_{i=1}^{m}|a_i|^q\right)^{1/q}$$

for some universal constant K (as long as $\varepsilon > 1/\sqrt{n}$), we get

$$(*)\quad \mu\{x; |f_{\bar{a}}(\bar{x}) - |||\bar{a}|||| < (K+1)\cdot L\cdot\varepsilon\cdot m^{\frac{1}{2}-\frac{1}{q}}\cdot\left(\textstyle\sum_{i=1}^{m}|a_i|^q\right)^{1/q}\} \geq 1-\sqrt{\pi/2}\ exp(-\varepsilon^2 n/2).$$

10.3.3. Using the antipodal property of φ and the cotype inequality we get,

$$|||\bar{a}||| = \int_X\int_0^1\left\|\sum_{i=1}^{m}a_i r_i(t)\varphi(x_i)\right\|dt\,d\mu(x) \geq C_q^{-1}\left(\sum_{i=1}^{m}|a_i|^q\right)^{1/q}.$$

Fix $\theta > 0$ and let $\varepsilon = \theta\{(K+1)\cdot L\cdot C_q\}^{-1}\cdot m^{\frac{1}{q}-\frac{1}{2}}$. Then

$$(K+1)\cdot L\cdot\varepsilon\cdot m^{\frac{1}{2}-\frac{1}{q}}\cdot\left(\sum_{i=1}^{m}|a_i|^q\right) \leq \theta\cdot|||\bar{a}|||$$

and we get

10.3.4.

$$\mu\{\bar{x}; |f_{\bar{a}}(\bar{x}) - |||\bar{a}|||| < \theta\cdot|||\bar{a}|||\} \geq 1-\sqrt{\pi/2}\ exp\left(\frac{-\theta^2\cdot n}{[m^{1-(2/q)}\cdot 2\cdot\{(K+1)\cdot L\cdot C_q\}^2]}\right).$$

Now pick a δ-net N in the sphere of the norm $|||\cdot|||$. One can do it with $|N| \leq e^{2m/\delta}$ (see Lemma 2.6). Therefore, if

$$\textbf{10.3.5.}\qquad \sqrt{\pi/2}\ exp\left(\frac{2m}{\delta} - \frac{\theta^2\cdot n}{m^{1-(2/q)}\cdot 2\cdot\{(K+1)\cdot L\cdot C_q\}^2}\right) < 1\,,$$

then one can find $\bar{x}\in X$ such that

$$(1-\theta)\cdot|||\bar{a}||| \leq \left\|\sum_{i=1}^{m}a_i\varphi(x_i)\right\| \leq (1+\theta)\cdot|||\bar{a}|||\ \ for\ all\ \ \bar{a}\in N.$$

It follows (as in 4.1 for example) that

$$\frac{1-2\delta-\theta}{1-\delta}\cdot|||\bar{a}||| \leq \left\|\sum_{i=1}^{m}a_i\varphi(x_i)\right\| \leq \frac{1+\theta}{1-\delta}\cdot|||\bar{a}|||$$

for all $\bar{a} \in \mathbb{R}^m$. In particular, $\{\varphi(x_i)\}_{i=1}^m$ is $(1+\theta)/(1-2\delta-\theta)$-symmetric. Choosing θ and δ of the same order and such that $(1+\theta)/(1-\theta-2\delta) = 1+\eta$ gives, using 10.3.5., the right estimate for m.

□

10.4. For the next result which is similar in nature to the previous one, we need the notion of a block sequence: Given a sequence $x_1, \ldots, x_n$ of elements of some linear space, any sequence of non- zero vectors of the form $\Sigma_{i\in\sigma_j} a_i x_i$, $j = 1, \ldots, m$, with $\{\sigma_j\}_{i=1}^m$ disjoint subsets of $\{1, \ldots, n\}$, is called a *block sequence* of $x_1, \ldots, x_n$.

THEOREM: *For any $\varepsilon > 0$ there exists a constant $c(\varepsilon) > 0$ such that, given any n and a sequence $x_1, \ldots, x_n$ of linearly independent vectors in a normed space E, with cotype q constant C_q, there exists a block sequence $\{y_1, \ldots, y_m\}$ of $x_1, \ldots, x_n$ which is $(1+\varepsilon)$-unconditional and $m \geq c(\varepsilon) n^{1/q}/C_q$.*

PROOF: We give only the beginning of the proof leaving the rest to the reader. Divide $\{1, \ldots, n\}$ into m almost equal parts $\{A_i\}_{i=1}^m$, $|A_i| \approx k \approx n/m$. In each $E_i = span\{x_j\}_{j\in A_i}$ find, using Theorem 9.6, a $(1+\frac{\varepsilon}{2})$-isomorphic copy F_i of ℓ_2^t for $t \geq c(\varepsilon) k^{2/q}/C_q$. Define now

$$\varphi : \Pi_{i=1}^m S(F_i) \to E$$

by $\varphi(x_1, \ldots, x_n) = \Sigma_{i=1}^m x_i$ and proceed as in 10.3.

□

10.5. *Finite dimensional version of Krivine's Theorem.* In Chapter 12 below we shall state and prove a theorem of Krivine about embedability of ℓ_p^n spaces in general Banach spaces. Here we present a special case of this theorem with a proof which, unlike the proof of the general theorem given in Chapter 12, allows one to get reasonable estimates on the dimension of the embedded ℓ_p^n space.

Let $\{e_i\}_{i=1}^n$ be the unit vector basis in $\mathbb{R}^n$ and let $\|\cdot\|_p$ be the ℓ_p norm on $\mathbb{R}^n$ (i.e., $\|\Sigma_{i=1}^n a_i e_i\|_p = (\Sigma_{i=1}^n |a_i|^p)^{1/p}$). Let $\|\cdot\|$ be another norm on $\mathbb{R}^n$ and assume

$$(*) \qquad C_1^{-1} \cdot \|x\|_p \leq \|x\| \leq C_2 \cdot \|x\|_p .$$

Put $C = C_1 \cdot C_2$.

THEOREM: *For any $\varepsilon > 0$ and $C < \infty$ there is a constant $\alpha = \alpha(\varepsilon, C)$ such that for any norm $\|\cdot\|$ on $\mathbb{R}^n$ satisfying $(*)$ there exists a block sequence $u_1 \ldots, u_m$ of $e_1 \ldots, e_n$ which satisfies*

$$(1-\varepsilon)\left(\sum_{i=1}^m |a_i|^p\right)^{1/p} \leq \left\|\sum_{i=1}^m a_i u_i\right\| \leq (1+\varepsilon)\left(\sum_{i=1}^m |a_i|^p\right)^{1/p}$$

for all $a_1, \ldots, a_m$ and $m \geq \alpha \cdot n^{\frac{1}{3}(\varepsilon/36C)^p}$.

The proof consists of two parts. In the first we find a symmetric block sequence

10.6. LEMMA: *For any $\varepsilon > 0$ there exists a constant $c(\varepsilon) > 0$ such that if $\|\cdot\|$ is a norm on $\mathbb{R}^n$ satisfying $(*)$ then there exists a block sequence $\{y_1, \ldots, y_k\}$ of $\{e_1, \ldots, e_n\}$ which is $(1+\varepsilon)$- symmetric and $k \geq c(\varepsilon) \cdot C^{-\frac{2p}{3}} \cdot n^{1/3}$.*

PROOF: Consider the family $\{X_n = E_2^n \times \Pi_n\}_{n=1}^{\infty}$ with the metric

$$\rho\{(t,\pi),(s,\varphi)\} = \frac{1}{n}\left|\{i;\ t_i \neq s_i \text{ or } \pi(i) \neq \varphi(i)\}\right|.$$

Using the natural sequence of σ-fields (an atom in the k-th σ-field is specified by the values of the first k components of $t \in E_2^n$ and the first k values, $\pi(1), \ldots, \pi(k)$, of $\pi \in \Pi_n$) it is easily seen that the length (see 7.7) of X_n is $\leq 2/\sqrt{n}$. By Theorem 7.8 we get that X_n is a normal Levy family (with constants $c_1 = 2$ and $c_2 = 1/64$).

Without loss of generality, we may assume that $n = k \cdot m$ (k to be chosen later). Divide $\{1, \cdots, n\}$ into k disjoint sets $A_1, \ldots, A_k$ each of cardinality m. For $(t,\pi) \in X_n$ let

$$y_j = y_j(t,\pi) = m^{-1/p} \sum_{i \in A_j} t_i e_{\pi(i)}, \quad j = 1, \ldots, k.$$

For $\bar{a} = (a_1, \ldots, a_k) \in \mathbb{R}^k$, consider the function

$$f_{\bar{a}}(t,\pi) = \left\|\sum_{j=1}^{k} a_j y_j\right\|$$

on X_n. We now estimate the modulus of continuity of $f_{\bar{a}}$

$$\omega_{\bar{a}}(\varepsilon) = m^{-1/p} \cdot \sup\left\{\left|\left\|\sum_{j=1}^{k} a_j \sum_{i \in A_j} t_i e_{\pi(i)}\right\| - \left\|\sum_{j=1}^{k} a_j \sum_{i \in A_j} s_i e_{\varphi(i)}\right\|\right|;\ \rho((t,\pi),(s,\varphi)) \leq \varepsilon\right\}$$

$$\leq C_2 \cdot m^{-1/p} \cdot \sup\left\{\left\|\sum_{j=1}^{k} a_j \sum_{i \in A_j} (t_i e_{\pi(i)} - s_i e_{\varphi(i)})\right\|_p;\ \rho\{(t,\pi),(s,\varphi)\} \leq \varepsilon\right\}.$$

For each j the maximal number of $i-s$ such that $t_i e_{\pi(i)} \neq s_i e_{\varphi(i)}$ is $\varepsilon \cdot n$.

Thus, we get,

$$\omega_{\bar{a}}(\varepsilon) \leq 2 \cdot C_2 \cdot m^{-1/p} \cdot (\varepsilon \cdot n)^{1/p} \cdot \left(\sum_{j=1}^{k} |a_j|^p\right)^{1/p}.$$

Clearly, $E(f_{\bar{a}}) = |||\bar{a}|||$ is a 1-symmetric norm on $\mathbb{R}^k$. The same computation as in 10.3.2 now yields

$$\mu\{(t,\pi);\ |f_{\bar{a}}(t,\pi) - |||\bar{a}||| | \leq K \cdot C_2(\varepsilon n/m)^{1/p}\|\bar{a}\|_p\} \geq 1 - 2\ exp(-\varepsilon^2 \cdot n/64).$$

Now

$$|||\bar{a}||| = E(f_{\bar{a}}) \geq C_1^{-1}||\bar{a}||^p,$$

plugging this into the inequality above yields, as in 10.3.4,

$$\mu\{(t,\pi);\ (1-\theta)|||\bar{a}||| \leq \left\|\sum_{j=1}^{k} a_j y_j(t,\pi)\right\| \leq (1+\theta)|||\bar{a}|||\} \geq 1 - 2\, exp\left[\frac{-m^2}{64n}\left(\frac{\theta}{2C}\right)^{2p}\right].$$

The rest of the proof follows steps 10.3.5 - 10.3.6 in the proof of Theorem 10.3 and, we hope, can be completed by the reader.

□

10.7. When proving Theorem 10.5 we may now assume, using Lemma 10.6, that we are dealing with a 1-symmetric sequence. The next theorem will then complete the proof of Theorem 10.5.

THEOREM: *If $y_1, \ldots, y_k$ is a 1-symmetric sequence in a normed space satisfying*

$$C_1^{-1}||\bar{a}||_p \leq \left\|\sum_{i=1}^{k} a_i y_i\right\| \leq C_2||\bar{a}||p$$

for all $\bar{a} = (a_1, \ldots, a_k) \in I\!R^k$. Then, for all $\varepsilon > 0$, there is a block sequence $u_1, \ldots, u_m$ of $y_1, \ldots, y_k$ satisfying

$$(1-\varepsilon)\cdot||\bar{a}||_p \leq \left\|\sum_{i=1}^{m} a_i u_i\right\| \leq (1+\varepsilon)\cdot||\bar{a}||_p$$

for all $\bar{a} = (a_1, \ldots . a_m) \in I\!R^m$ where $m \geq \Gamma^{3p} k^{\Gamma}$, $\Gamma = (\varepsilon/36C)^p$, $C = C_1 \cdot C_2$.

PROOF: After nine chapters of results in which we found nice substructures of given structures, using probabilistic method, it is an interesting feature of the following proof that it is deterministic. One can actually write a formula (depending on ε, p and k) for the block sequence $\{u_i\}_{i=1}^m$. We begin with a description of this formula.

Let "$\cong$" mean "is a rearrangement of" and "$\oplus$" mean "sum with disjoint supports". Let Y_r stand for the $\cong$ equivalence class of the sum of r distinct elements of $\{y_1, \ldots, y_k\}$. Let z_s, $s = 1, \ldots, m$, be disjoint with

$$z_s \cong \sum_{j=0}^{N-1} \oplus \rho^{(N-j)/p} \cdot Y_{[\rho^j]}$$

where $\rho = (a+1)/a$ and the integers a, N, depending on ε, p and k, will be chosen later. The integer m is determined by

$$(*)\qquad k \approx m \cdot \sum_{j=0}^{N-1} [\rho^j] \approx m \cdot \sum_{j=0}^{N-1} \rho^j \leq m\rho^N/(\rho-1) = ma((a+1)/a)^N).$$

Finally we set

$$u_s = z_s/\|z_s\|, \qquad s = 1, \dots, m.$$

The dependence of a and N on ε, p and k may be traced through the following formulas which we bring in detail in order that the dependence of the different parameters in the proof will be clear:

$$\varepsilon_0 = \frac{1}{12}(\varepsilon/2C)^p, \quad \delta = \varepsilon_0 \cdot m^{-1/p}, \quad a = [1/\delta p] = [m^{1/p}/p\varepsilon_0], (a/(a+1))^{[N\varepsilon_0]} \sim \varepsilon_0^p/m.$$

The substitution of all the relations above in (*) will determine m.

We now give a formal proof. Fix $T < N$, $M \leq m$ and $t_s \in \{0, 1, \dots, T\}$, $s = 1, \dots, M$, with

$$\sum_{s=1}^{M} \rho^{-t_s} = 1 + \eta, \qquad |\eta| \leq 1.$$

Then

$$\sum_{s=1}^{M} \rho^{-t_s/p} z_s \cong \sum_{\substack{s=1,\dots,M \\ j=0,\dots,N-1}} \oplus\, \rho^{(N-j-t_s)/p} Y_{[\rho^j]} \cong \sum_{i=0}^{N-1+T} \oplus\, \rho^{(N-i)/p} Y_{a_i},$$

where

$$a_i = \sum_{\substack{s \in \{1,\dots,M\} \\ i-t_s \in \{0,\dots,N-1\}}} [\rho^{i-t_s}].$$

Observe that (for $i \in \{0, \dots, N-1\}$)

$$\begin{aligned}[\rho^i] - a_i &= \left(\sum_{s=1}^{M} \rho^{-t_s} - \eta\right)[\rho^i] - a_i \\ &= \sum_{i-t_s<0} \rho^{-t_s}[\rho^i] + \sum_{i-t_s \geq 0} ([\rho^i]\rho^{-t_s} - [\rho^{i-t_s}]) - \eta[\rho^i].\end{aligned}$$

Comparing $\Sigma_{s=1}^{M} \rho^{-t_s/p} z_s$, with the appropriate rearrangement z of z_1, the difference Δ satisfies

$$\Delta \cong \sum_{i=0}^{N-1} \rho^{(N-i)/p} Y_{|[\rho^i]-a_i|} \oplus \sum_{i=N}^{N-1+T} \oplus\, \rho^{(N-i)/\rho} Y_{a_i}.$$

Hence

$$\|\Delta\| \le C_2\left(\sum_{i=0}^{N-1}\rho^{N-i}\big|[\rho^i]-a_i\big| + \sum_{i=N}^{N-1+T}\rho^{N-i}a_i\right)^{1/p}$$

$$\le C_2\rho^{N/p}\left(\sum_{i=0}^{N-1}\left(\sum_{i-t_s<0}\rho^{-i-t_s}[\rho^i]\right.\right.$$

$$\left.+\sum_{i-t_s\ge 0}\big|[\rho^i]\rho^{-i-t_s}-\rho^{-i}[\rho^{i-t_s}]\big|+|\eta|[\rho^i]\rho^{-i}\right)$$

$$\left.+\sum_{i=N}^{N-1+T}\sum_{i-t_s<N}\rho^{-i}[\rho^{i-t_s}]\right)^{1/p}$$

$$\le C_2\rho^{N/p}\left(\sum_{i=0}^{T-1}\left(\sum_{i-t_s<0}\rho^{-t_s}+\sum_{i-t_s\ge 0}\rho^{-i}+|\eta|\right)\right.$$

$$\left.+\sum_{i=T}^{N-1}\left(\sum_{s=1}^{M}\rho^{-i}+|\eta|\right)+\sum_{i=N}^{N-1+T}\sum_{i-t_s<N}\rho^{-t_s}\right)^{1/p}$$

$$\le C_2\rho^{N/p}(2(1+|\eta|)T+N|\eta|)T+N|\eta|+(N-T)M\rho^{-T})^{1/p}$$

$$\le C_2\rho^{N/p}(4T+N|\eta|+NM\rho^{-T})^{1/p}.$$

On the other hand, $\|z_s\|\ge(1/C_1)\rho^{N/p}(N/2)^{1/p}$, hence the relative difference is

$$\left|\left\|\sum_{s=1}^{k}\rho^{-t_s/p}u_s\right\|-1\right|\le\|\Delta\|/\|z_s\|\le C\big(8T/N+2|\eta|+2M\rho^{-T}\big)^{1/p},$$

and this is $\le\varepsilon/2$, provided $T\le N\varepsilon_0$, $|\eta|\le\varepsilon_0$ and $M\rho^{-T}\le\varepsilon_0$, where $\varepsilon_0=\frac{1}{12}(\varepsilon/2C)^p$. We assume, therefore, $T=[N\varepsilon_0]$. Let now

$$\sum_{s=1}^{m}|\alpha_s|^p=1.$$

We want to replace the α_s by some $\beta_s=\rho^{-t_s/p}$, $t_s\in\{0,\dots,T\}$ or by $\beta_s=0$, so that $\big||\alpha_s|-\beta_s\big|\le\delta$ $\forall s$ where $\delta=\varepsilon_0 m^{-1/p}$. For this we ask that $\rho^{-T/p}\le\delta$, i.e., $(a/(a+1))^{[N\varepsilon_0]}\le\varepsilon_0^p/m$ and that $1-\rho^{-1/p}\le\delta$. Taking $a=[1/\delta p]=m^{1/p}/p\varepsilon_0$ guarantees this. The choice of δ implies that

$$\left|\left(\sum_{s=1}^{m}\beta_s^p\right)^{1/p}-1\right|\le\varepsilon_0,$$

and that

$$\left|\left\|\sum_{s=1}^{m}\beta_su_s\right\|-\left\|\sum_{s=1}^{m}\alpha_su_s\right\|\right|\le\left\|\sum_{s=1}^{m}\big|\beta_s-|\alpha_s|\big|u_s\right\|\le\delta\left\|\sum_{s=1}^{m}u_s\right\|\le C\varepsilon_0<\frac{\varepsilon}{2}.$$

Hence $|\|\sum_{s=1}^{m} \alpha_s u_s\| - 1| < \varepsilon$. The conditions which have to be guaranteed are thus, $m \leq \varepsilon_0^p((a+1)/a)^{[N\varepsilon_0]}$ and $ma((a+1)/a)^N \leq k$, when $a = [m^{1/p}/\varepsilon_0 p]$. The substitution

$$\left(\frac{a}{a+1}\right)^{[N\varepsilon_0]/p} \sim \varepsilon_0 m^{-1/p}$$

yields

$$m\frac{m^{1/p}}{\varepsilon_0 p}\left(\frac{m^{1/p}}{\varepsilon_0}\right)^{p/\varepsilon_0} \sim k,$$

i.e.,

$$\begin{aligned} m &\sim \varepsilon_0^{p(1+\varepsilon_0/p)/(1+\varepsilon_0(1+1/p))}(pk)^{\varepsilon_0/(1+\varepsilon_0(1+1/p))} \\ &\geq \left(\frac{\varepsilon}{24C}\right)^{5p^2/2} k^{(\varepsilon/36C)^p} \geq \Gamma^{3p}k^{\Gamma} \end{aligned}$$

where $\Gamma = (\varepsilon/36C)^p$ (provided $\varepsilon < 6C$).

$\square$

PART II: TYPE AND COTYPE OF NORMED SPACES

11. RAMSEY'S THEOREM WITH SOME APPLICATIONS TO NORMED SPACES

11.1. We begin with the statement and proof of Ramsey's Theorem. For a set B and a positive integer k let $B^{[k]}$ denote the set of all subsets of B with k elements.

THEOREM: *Let A be a finite set, k a positive integer and let $f : \mathbb{N}^{[k]} \to A$ be any function. Then there exists an infinite subset $M \subseteq \mathbb{N}$ such that $f(M^{[k]})$ is a singleton.*

A simple compactness argument shows that the following finite version of the theorem also holds.

11.2. THEOREM: *For all positive integers k, n, m there exists an N such that if $f : \{1,\ldots,N\}^{[k]} \to \{1,\ldots,m\}$ is any function then there exists a subset $M \subseteq \{1,\ldots,N\}$ with $|M| = n$ such that $f(M^{[k]})$ is a singleton.*

The behaviour of N as a function of k, n, m is a much studied subject (cf. [G.R.S.]).

11.3. PROOF OF 11.1: We shall prove the Theorem by induction on k. The case $k = 1$ is evident. Assume the Theorem holds for k and let $f : \mathbb{N}^{[k+1]} \to A$ be any function from the subsets of $\mathbb{N}$ of cardinality $k+1$ to a finite set A. Let $m_1 \in \mathbb{N}$ be arbitrary and consider the function $f_{m_1} : (\mathbb{N} - \{m_1\})^{[k]} \to A$ given by

$$f_{m_1}(\{n_1,\ldots,n_k\}) = f(\{m_1,n_1,\ldots,n_k\}) .$$

By the induction hypothesis there exists an infinite $M_1 \subseteq \mathbb{N} - \{m_1\}$ such that $f(\{m_1,n_1,\ldots,n_k\})$ is a constant for all $\{n_1 \ \ldots \ n_k\} \subseteq M_1$. Let $m_2 \in M_1$ and choose an infinite $M_2 \subseteq M_1 - \{m_2\}$ such that $f(\{m_2,n_1,\ldots,n_k\})$ is a constant for all $\{n_1,\ldots,n_k\} \subseteq M_2$. Continuing in the same manner we find a sequence of infinite subsets $\mathbb{N} = M_0 \supseteq M_1 \supseteq M_2 \supseteq M_3 \supseteq \ldots$ of the integers and a sequence $m_1, m_2, \ldots$ with $m_i \in M_{i-1} - M_i$, $i = 1,2,\ldots$, and

$$f(\{m_i,n_1,\ldots,n_k\}) \quad = \quad \text{const} \qquad (*)$$

for all $\{n_1,\ldots,n_k\} \subseteq M_i$.

Choose an infinite subsequence $M \subseteq \{m_1, m_2, \ldots\}$ such that the constant in $(*)$ is the same for all the members of M. M is then the desired subsequence.

□

11.4. We turn now to the first application to normed spaces.

THEOREM: *Let $\{x_i\}_{i=1}^{\infty}$ be a sequence in a normed space X with the property that for all n and all scalars $a_1, \ldots, a_n$ not all of which are zero, there exist constants $0 < c < C < \infty$ (depending on $a_1, \ldots, a_n$) such that for all $i_1 < \ldots < i_n$*

$$c \leq \left\| \sum_{j=1}^{n} a_j x_{i_j} \right\| \leq C .$$

Then given any sequence $\varepsilon_1 > \varepsilon_2 > \ldots > 0$ there exists an infinite subsequence $M = \{m_n\} \subseteq \mathbb{N}$ such that for all n, all $m_n < i_1 < \ldots < i_n$, $m_n < j_1 < \ldots < j_n$, $i_k, j_k \in M$ and all scalars $a_1, \ldots, a_n$

$$\left\| \sum_{k=1}^{n} a_k x_{i_k} \right\| \leq (1 + \varepsilon_n) \left\| \sum_{k=1}^{n} a_k x_{j_k} \right\| .$$

PROOF: We build infinite sequences $N_1 \supseteq N_2 \supseteq \ldots$ of integers such that the assertion holds for N_n for a fixed n. M will then be the diagonal sequence.

To build N_n given N_{n-1} we proceed as follows: Fix $a_1, \ldots, a_n$, divide the interval $[c, C]$ into finitely many disjoint intervals of length at most $\delta > 0$ (to be chosen later) and let $f : N_{n-1}^{[n]}$ be the function which assigns to $i_1, \ldots, i_n$, satisfying $i_1 < \ldots < i_n$, the number of the interval to which $\|\Sigma_{j=1}^{n} a_j x_{i_j}\|$ belongs. By Ramsey's Theorem 11.1 one can find a subsequence M_1 such that $\|\Sigma_{j=1}^{n} a_j x_{i_j}\|$ always belongs to the same interval for $i_1, \ldots, i_n \in M_1$. If $\delta > 0$ is small enough, this ensures that

$$\left\| \sum_{k=1}^{n} a_k x_{i_k} \right\| \leq \left(1 + \frac{\varepsilon_n}{2}\right) \left\| \sum_{k=1}^{n} a_k x_{j_k} \right\|$$

for any two sequences $i_1 < \ldots < i_n$, $j_1 < \ldots < j_n$ in M_1. We now repeat the process for a different sequence $a_1, \ldots, a_n$ to find a subsequence M_2 of M_1 good also for these scalars. We repeat the process finitely many times to take care of all scalars $(a_1, \ldots, a_n)$ in a fine enough net of $B(\ell_1^n)$. The last sequence will be N_n. We leave the details to the reader.

□

11.5. COROLLARY: *Let M be the sequence given in Theorem 11.4. Then the sequence*

$$y_i = x_{m_{2i}} - x_{m_{2i-1}} \ , i = 1, 2, \ldots$$

satisfies

$$\left\| \sum_{i \in A} a_i y_i \right\| \leq (1 + \varepsilon_n) \left\| \sum_{i \in B} a_i y_i \right\|$$

for any finite sets $A \subseteq B$ with $\min B \geq |B|$.

PROOF: Fix a finite set B and scalars $\{a_i\}_{i\in B}$. Fix $n \in \mathbb{N}$ and let $\{\sigma_i\}_{i\in B}$ be subsets of M with the following properties:

(i) if $i > j$ then $\min \sigma_i > \max \sigma_j$

(ii) $\min \sigma_i > m_{|B|\cdot(n+1)}$ for any $i \in B$

(iii) for $i \in A$ $|\sigma_i| = 2$. Say, $\sigma_i = \{j(i,1), j(i,2)\}$

(iv) for $i \in B - A$ $|\sigma_i| = n+1$. Say, $\sigma_i = \{j(i,1), \dots, j(i,n+1)\}$.

By the property of M (almost invariance under spreading) we get for any $1 \leq k \leq n$

$$\left\|\sum_{i\in A} a_i(x_{j(i,2)} - x_{j(i,1)}) + \sum_{i\in B-A} a_i(x_{j(i,k+1)} - x_{j(i,k)})\right\| \leq (1+\varepsilon_n)\left\|\sum_{i\in B} a_i y_i\right\| .$$

Averaging over $1 \leq k \leq n$, we get

$$\begin{aligned}(1+\varepsilon_n)\left\|\sum_{i\in B} a_i y_i\right\| &\geq \frac{1}{n}\left\|n\sum_{i\in A} a_i(x_{j(i,2)} - x_{j(i,1)}) + \sum_{k=1}^{n}\sum_{i\in B-A} a_i(x_{j(i,k+1)} - x_{j(i,k)})\right\| \\ &= \left\|\sum_{i\in A} a_i(x_{j(i,2)} - x_{j(i,1)}) + \frac{1}{n}\sum_{i\in B-A} a_i(x_{j(i,n+1)} - x_{j(i,1)})\right\| .\end{aligned}$$

Letting $n \to \infty$ we get the desired result.

□

11.6. DEFINITION: We say that a Banach space Y is ***K-finitely representable*** in a Banach space X if for every finite dimensional subspace E of Y there exists an isomorphism $T : E \xrightarrow{into} X$ with $\|T\| \cdot \|T^{-1}\| \leq K$. Y is said to be ***crudely finitely representable*** in X if it is K-finitely representable in X for *some* K. It is said to be ***finitely representable*** in X if it is K finitely representable in X for *all* $K > 1$.

Using this definition one can state, for example, Dvoretzky's theorem as stating that ℓ_2 is finitely representable in any infinite dimensional normed space.

DEFINITION: We say that a sequence $\{y_n\}_{n=1}^{\infty}$ in a Banach space Y is ***K-block finitely representable*** on a sequence $\{x_n\}_{n=1}^{\infty}$ in a Banach space X if for each N there exists N vectors $\{z_n\}_{n=1}^{N}$ in X of the form $z_n = \Sigma_{i\in\sigma_n} a_i x_i$, σ_n disjoint, such that

$$\left\|\sum_{n=1}^{N} \alpha_n y_n\right\| \leq \left\|\sum_{n=1}^{N} \alpha_n z_n\right\| \leq K\left\|\sum_{n=1}^{N} \alpha_n y_n\right\|$$

for all choices of $\{\alpha_n\}_{n=1}^{N}$.

We define now the notions of ***crudely block finite representability*** and ***block finite representability*** in an analogous way to that of the previous definition.

11.7. COROLLARY: *Let $\{x_i\}_{i=1}^{\infty}$ be any sequence satisfying the assumption of Theorem 11.4. Then there exists a sequence $\{\overline{y}_i\}_{i=1}^{\infty}$ (of nonzero vectors) in some Banach space Y which is block finitely representable on $\{x_i\}_{i=1}^{\infty}$ and have the two properties.*

(i) It is invariant under spreading, i.e.

$$\left\|\sum_{i=1}^{n} a_i \overline{y}_i\right\| = \left\|\sum_{i=1}^{n} a_i \overline{y}_{k_i}\right\|$$

for all $n, k_1 < \ldots < k_n$ and all scalars $a_1, \ldots, a_n$.

(ii) It satisfies

$$\left\|\sum_{i\in A} a_i \overline{y}_i\right\| \leq \left\|\sum_{i\in B} a_i \overline{y}_i\right\|$$

for all $A \subseteq B$, B finite, and all scalars $\{a_i\}_{i\in B}$.

PROOF: Let Y be the completion of the space of all finite sequences endowed with the norm

$$\|\{a_i\}_{i=1}^{n}\| = \lim_{k\to\infty}\left\|\sum_{i=1}^{n} a_i y_{k+i}\right\|$$

where $\{y_i\}_{i=1}^{\infty}$ is as in Corollary 11.5 (with $\varepsilon_n \to 0$) and let $\{\overline{y}_i\}_{i=1}^{\infty}$ be the unit vector basis in this space.

□

11.8. Note that (ii) of Corollary 11.7 implies

$$\left\|\sum_{i=1}^{n} a_i \overline{y}_i\right\| \leq 2\left\|\sum_{i=1}^{n} \pm a_i \overline{y}_i\right\|$$

for all n, $a_1, \ldots, a_n$ and all choices of signs. One can actually strengthen this to get the constant 1.

THEOREM: *Let $\{x_i\}_{i=1}^{\infty}$ be a sequence as in Theorem 11.4. Then there exists a sequence $\{z_i\}_{i=1}^{\infty}$ of nonzero vectors in some Banach space Y which is block finitely representable on $\{x_i\}_{i=1}^{\infty}$ and has the two properties.*

(i) It is 1 - spreading

$$\left\|\sum_{i=1}^{n} a_i z_i\right\| = \left\|\sum_{i=1}^{n} a_i z_{k_i}\right\|$$

for all $n, k_1 < \ldots < k_n$ and all scalars $a_1, \ldots, a_n$ and

(ii) It is 1 - unconditional;

$$\left\|\sum_{i=1}^{n} a_i z_i\right\| = \left\|\sum_{i=1}^{n} \pm a_i z_i\right\|$$

for all n and $a_1, \ldots, a_n$.

PROOF: Since block finite representability is a transitive property it is enough to prove the theorem for $\{x_i\} = \{\overline{y}_i\}$ of the previous corollary. We may also assume $\|\overline{y}_i\| = 1$. We shall distinguish between two cases:

Case 1:

$$\left\|\sum_{i=1}^{n} \overline{y}_i\right\| \underset{n\to\infty}{\longrightarrow} \infty .$$

Fix k and n. Let $A_1, \ldots, A_n$ be intervals of integers each of length $2k$, $A_i = (j_i, j_i + 1, \ldots, j_i + 2k - 1)$ with $j_i + 2k < j_{i+1}$ for all $i = 1, \ldots, n$. Let

$$u_i = \overline{y}_{j_i} - \overline{y}_{j_i+1} + \overline{y}_{j_i+2} - \cdots - \overline{y}_{j_i+2k-1} .$$

Then

$$\|u_i\| \geq \frac{1}{2}\left\|\sum_{i=1}^{2k} \overline{y}_i\right\| \underset{k\to\infty}{\longrightarrow} \infty .$$

Put, for $i = 1, \ldots, n$,

$$v_i = u_i/|u_i| .$$

Given $a_1, \ldots, a_n$ and signs $\varepsilon_1, \ldots, \varepsilon_n$ let T_ε be the isometry on span $\{y_i\}_{i=1}^{\infty}$ given by

$$T_\varepsilon y_j = y_j \quad \text{if} \quad j \in A_i \quad \text{and} \quad \varepsilon_i = 1$$
$$T_\varepsilon y_j = y_{j+1} \quad \text{if} \quad j \in A_i \quad \text{and} \quad \varepsilon_i = -1 .$$

Notice that for i such that $\varepsilon_i = 1$, $T_\varepsilon v_i = v_i$ and for i such that $\varepsilon_i = -1$

$$T_\varepsilon v_i - v_i = \|u_1\|^{-1}\|y_1 - y_2\| .$$

So

$$\|T_\varepsilon(\sum_{i=1}^{n} a_i v_i) - \sum_{i=1}^{n} a_i v_i\| \leq 2\|u_1\|^{-1} \sum_{i=1}^{n} |a_i| \leq 2\|u_1\|^{-1} \cdot n \cdot \max_{1\leq i\leq n} |a_i|$$
$$\leq 2\|u_1\|^{-1} \cdot n \cdot \left\|\sum_{i=1}^{n} a_i v_i\right\| .$$

If k is large enough with respect to n we get

$$\left\|T_\varepsilon(\sum_{i=1}^{n} a_i v_i) - \sum_{i=1}^{n} a_i v_i\right\| \leq \frac{1}{n}\left\|\sum_{i=1}^{n} a_i v_i\right\|$$

and in particular

$$(1 - \frac{1}{n})\left\|\sum_{i=1}^{n} a_i v_i\right\| \leq \left\|T_\varepsilon(\sum_{i=1}^{n} a_i v_i)\right\| \leq (1 + \frac{1}{n})\left\|\sum_{i=1}^{n} a_i v_i\right\| .$$

Note also that

$$\left\|\sum_{i=1}^{n} a_i v_i\right\| \geq \left\|\sum_{i\in B} a_i v_i\right\|$$

for all $B \subseteq \{1, \dots, n\}$ (since this holds for the y_i's).

The v_i's depend on n, denote $v_i = v_i^n$. For any sequence $a_1, \dots, a_m$ of scalars the sequence $(\|\Sigma_{i=1}^m a_i v_i^n\|)_{n=1}^\infty$ is bounded. We may pass to a subsequence M such that

$$\|\{a_i\}_{i=1}^m\| = \lim_{n \in M} \left\| \sum_{i=1}^n a_i v_i^n \right\|$$

exists for a dense set of all finite sequences of scalars $\{a_i\}_{i=1}^m$ (and thus for all the finite sequences). The limit is necessarily a norm (note $\|\{a_i\}_{i=1}^m\| \geq max|a_i|$). The sequence $\{z_i\}_{i=1}^\infty$ can be taken to be the unit vector basis in the completion of the space of finite sequences with respect to this norm.

Case 2: There exists an M such that $\|\Sigma_{i=1}^n \overline{y}_i\| \leq M$ for all n. (Note: $\|\Sigma_{i=1}^n \overline{y}_i\|$ is a nondecreasing sequence so that the two cases are inclusive).

In this case we shall show that the unit vector basis of c_0, the space of all sequences tending to zero, with the sup norm, is block finitely representable on $\{\overline{y}_i\}_{i=1}^\infty$, i.e., that for each $\varepsilon > 0$ and $k > 0$ there are disjoint blocks $z_1, \dots, z_k$ of $\{\overline{y}_i\}_{i=1}^\infty$ such that

$$\left\| \sum_{i=1}^k a_i z_i \right\| \leq (1 + \varepsilon) \max_{1 \leq i \leq k} |a_i| \cdot \|z_i\|$$

(note that the property (i) of the $\{\overline{y}_i\}$ immediately implies $\|\Sigma_{i=1}^k a_i z_i\| \geq \max_{1 \leq i \leq k} |a_i| \cdot \|z_i\|$).

First note that

$$\left\| \sum_{i=1}^n a_i \overline{y}_i \right\| \leq 2M \max_{1 \leq i \leq n} |a_i| \, .$$

Indeed, let f be a norm one functional such that $\|\Sigma_{i=1}^n a_i \overline{y}_i\| = f(\Sigma_{i=1}^n a_i \overline{y}_i)$, then,

$$\begin{aligned} \left\| \sum_{i=1}^n a_i \overline{y}_i \right\| &= \sum_{i=1}^n |a_i| f((sign\ a_i)\overline{y}_i) \\ &\leq \max_{1 \leq i \leq n} |a_i| \cdot f(\sum_{i=1}^n (sign\ a_i)\overline{y}_i) \\ &\leq \max_{1 \leq i \leq n} |a_i| \cdot \| \sum_{i=1}^n (sign\ a_i)\overline{y}_i \| \leq 2M \max_{1 \leq i \leq n} |a_i| \, . \end{aligned}$$

The conclusion will now follow from the next lemma.

LEMMA: *Let $(u_i)_{i=1}^{n^2}$ be a sequence of nonzero vectors in a Banach space X with the property*

$$\left\| \sum_{i=1}^{n^2} a_i u_i \right\| \leq K \cdot \max_{1 \leq i \leq n^2} |a_i| \cdot \|u_i\|$$

for all $a_1, \dots, a_{n^2}$.

Then there exist n disjoint nonzero blocks $v_1, \ldots, v_n$ such that

$$\left\|\sum_{i=1}^{n} a_i v_i\right\| \le \sqrt{K} \cdot \max_{1\le i\le n} |a_i| \cdot \|v_i\|$$

for all $a_1, \ldots, a_n$.

PROOF: For $v = \Sigma a_i u_i$ we denote $\|v\|_\infty = max|a_i| \cdot \|u_i\|$. If, for some $1 \le k \le n$ and all $a_1, \ldots, a_n$,

$$\left\|\sum_{i=1}^{n} a_i u_{(k-1)n+i}\right\| \le \sqrt{K} \cdot \max_{1\le i\le n} |a_i| \cdot \|u_{(k-1)n+i}\|$$

we are done. Otherwise there exist $v_k \in span\{u_{(k-1)n+i}\}_{i=1}^{n}$, $k = 1, \ldots, n$, such that

$$\|v_k\| \ge \sqrt{K} \cdot \|v_k\|_\infty .$$

Then for all $a_1, \ldots, a_n$,

$$\begin{aligned}\left\|\sum_{k=1}^{n} a_k v_k\right\| &\le K \left\|\sum_{k=1}^{n} a_k v_k\right\|_\infty \\ &= K \max_{1\le k\le n} |a_k| \, \|v_k\|_\infty \le \sqrt{K} \cdot \max_{1\le k\le n} |a_k| \cdot \|v_k\| .\end{aligned}$$

□

11.9. We conclude this chapter with two propositions which are finite dimensional analogues of results from this chapter. We shall use them in Chapter 13 below.

PROPOSITION: *For every $\varepsilon > 0$ and positive integer k there exists an $N = N(k, \varepsilon)$ such that if $n > N$ and $y_1, \ldots, y_n$ is any sequence of norm one elements in some normed space then there exists a subsequence $y_{i_1}, \ldots, y_{i_k}$ which is either $(1+\varepsilon)$-spreading (i.e. $\|\Sigma_{j=1}^{t} a_j y_{i_{n_j}}\| \le (1+\varepsilon)\|\Sigma_{j=1}^{t} a_j y_{i_{m_j}}\|$ for any two subsequences $n_1 < n_2 < \ldots < n_t$, $m_1 < m_2 < \ldots < m_t$ of $1, \ldots, k$) or satisfies*

$$(1-\varepsilon)k^{\frac{1}{2}} \le \operatorname*{Ave}_{\pm} \left\|\sum_{j=1}^{k} \pm y_{i_j}\right\| \le (1+\varepsilon)k^{\frac{1}{2}} .$$

PROOF: If this is not the case, let k and ε be such that for any n sufficiently large there exists a sequence $y_1^n, \ldots y_n^n$ of norm one elements in some normed space for which the conclusion of the proposition does not hold with these k and ε. For an appropriate subsequence of the positive integers, M, the limits

$$\lim_{n\in M} \left\|\sum_{i=1}^{\ell} a_i y_i^n\right\|$$

exist for all finite sequences $a_1, \ldots, a_\ell$. Let $\{e_i\}_{i=1}^{\infty}$ be the natural basis in the space of all finite sequences and define

$$\left\|\sum_{i=1}^{\ell} a_i e_i\right\| = \lim_{n\in M} \left\|\sum_{i=1}^{\ell} a_i y_i^n\right\| .$$

This is a seminorm. Taking the quotient by the null space (i.e. identifying any two sequences for which the seminorm of their difference is zero) we get a normed space.

As in the proof of Theorem 11.4 one can find a subsequence, M_1, of the positive integers for which

$$\lim_{\substack{s_1<s_2<\ldots<s_k \\ s_i\in M_1, s_1\to\infty}} \left\| \sum_{i=1}^{k} a_i e_{s_i} \right\|$$

exists for every sequence $a_1, \ldots, a_k$. If for some $a_1, \ldots, a_n$ with $\Sigma_{i=1}^k |a_i| \neq 0$ this limit is zero then clearly,

$$\lim_{s_i, s_j \to \infty, s_i, s_j \in M_1} \| e_{s_i} - e_{s_j} \| = 0 \ .$$

We then easily get that, for some $s_1 < \ldots < s_k$ in M_1,

$$(1-\frac{\varepsilon}{2})k^{\frac{1}{2}} \leq \operatorname*{Ave}_{\pm} \left\| \sum_{i=1}^{k} \pm e_{s_i} \right\| \leq (1+\frac{\varepsilon}{2})k^{\frac{1}{2}}$$

and then that, if $n \in M$ is large enough, then

$$(1-\varepsilon)k^{\frac{1}{2}} \leq \operatorname*{Ave}_{\pm} \left\| \sum_{i=1}^{k} \pm y_{s_i}^n \right\| \leq (1+\varepsilon)k^{\frac{1}{2}} \ .$$

If

$$\lim_{\substack{s_1<\ldots<s_k \\ s_i\in M_1, s_1\to\infty}} \left\| \sum_{i=1}^{m} a_i e_{s_i} \right\| > 0$$

for all $a_1, \ldots, a_k$ with $\Sigma_{i=1}^k |a_i| \neq 0$, we get similarly a subsequence of length k in some set $\{y_1^n, \ldots, y_n^n\}$ which is $(1+\varepsilon)$-spreading. This contradiction proves the proposition.

□

11.10. The last proposition is a finite dimensional analogue of Corollary 11.5. The proof is similar to the proof of Proposition 11.9 and we omit it.

PROPOSITION: *Let $\delta, \varepsilon > 0$ and let n be a positive integer. there exists an $N = N(n, \delta, \varepsilon)$ such that if $x_1, \ldots, x_N$ are any N elements in the unit ball of a normed space satisfying $\|x_i - x_j\| \geq \delta$ for $i \neq j$. Then there exists a subsequence $x_{i_1}, \ldots, x_{i_{2n}}$ such that the sequence $v_j = x_{i_{2j}} - x_{i_{2j-1}}$, $j = 1, \ldots, n$ is $2+\varepsilon$ unconditional.*

12. KRIVINE'S THEOREM

The main object of this chapter is to prove Krivine's Theorem (12.4) stating that the unit vector basis of some ℓ_p, $1 \le p < \infty$ or c_0 is block finitely representable on any non degenerate sequences in a Banach space. For later use in the next chapter we also need a version identifying the right p. This is given in 12.5.

12.1. We begin by recalling the notions of approximate eigenvalue and approximate eigenvector. Let X be a Banach space over the complex field and let $T : X \to X$ be a bounded linear operator. A sequence $\{u_n\}_{n=1}^{\infty}$ of norm one vectors in X is called an *approximate eigenvector with approximate eigenvalue* λ ($\in \mathbb{C}$) if

$$\|Tu_n - \lambda u_n\| \longrightarrow 0 .$$

The next lemma shows that any operator has an approximate eigenvalue. We sketch the proof briefly.

LEMMA: *Let $T : X \to X$ be a bounded linear operator and let λ be any member of the boundary of the spectrum of $T, \lambda \in \partial\sigma(T)$. Then λ is an approximate eigenvalue.*

PROOF: Let $\lambda_n \in \mathbb{C}\setminus\sigma(T)$ be such that $\lambda_n \to \lambda$ then $\|(T - \lambda_n I)^{-1}\| \to \infty$. Otherwise $\|(T-\lambda_n I)^{-1}\|\cdot|\lambda_n - \lambda| < \frac{1}{2}$ for n sufficiently large and it would follow that $T - \lambda I$ is invertible (with $(T - \lambda I)^{-1} = (T - \lambda_n I)^{-1}\Sigma_{n=0}^{\infty}(I - (T - \lambda I)(T - \lambda_n I)^{-1})^n$).

- Let $v_n \in X$ be such that $\|v_n\| = 1$ and $\|(T - \lambda_n I)v_n\| \to 0$. Then also $\|(T - \lambda I)v_n\| \to 0$.

□

12.2. We now prove a similar assertion for two commuting operators.

PROPOSITION: *Let X be a complex Banach space and let $T, S : X \to X$ be two bounded linear operators with $TS = ST$. Let λ be any approximate eigenvalue of T. Then there exists $\mu \in \sigma(S)$ and a sequence $\{u_n\}_{n=1}^{\infty}$ of norm one vectors in X such that*

$$\|Tu_n - \lambda u_n\| \underset{n\to\infty}{\longrightarrow} 0 \quad \textit{and} \quad \|Su_n - \lambda u_n\| \underset{n\to\infty}{\longrightarrow} 0 .$$

PROOF: Let $\underset{i\to\infty}{\text{LIM}}\, a_i$ denote a Banach limit over bounded sequences, i.e. $\underset{i\to\infty}{\text{LIM}}$ is a norm one, positive, linear functional over ℓ_∞, the space of bounded sequences with the sup norm, with the two properties:

(i) $LIM_{i\to\infty} a_i = LIM_{i\to\infty} a_{i+1}$ and

(ii) $LIM_{i\to\infty} a_i = lim_{i\to\infty} a_i$ for all convergent sequences $\{a_i\}_{i=1}^{\infty}$. (See [D.S.], p.73).
Define a seminorm on $(\Sigma \oplus X)_\infty$ by,

$$\|\{x_i\}_{i=1}^{\infty}\| = \underset{i\to\infty}{\text{LIM}} \|x_i\| \quad \text{for} \quad \{x_i\}_{i=1}^{\infty} \in (\Sigma \oplus X)_\infty .$$

Let N be the null space of this seminorm and let

$$\overline{X} = (\Sigma \oplus X)_\infty / N \ .$$

Define $\overline{T} : \overline{X} \to \overline{X}$ by $\overline{T}(x_1, x_2, \ldots) = (Tx_1, Tx_2, \ldots)$ and define similarly $\overline{S}$. Note that for $\overline{v} = (v_1, v_2, \ldots)$, $\overline{T}\overline{v} = \lambda \overline{v}$ so that λ is an eigenvalue of $\overline{T}$. Let $Y \subseteq \overline{X}$ be the space of all eigenvectors of $\overline{T}$ with eigenvalue λ. Since $\overline{T}\,\overline{S} = \overline{S}\,\overline{T}$, $\overline{S}Y \subseteq Y$. Let $\mu \in \partial\sigma(\overline{S}|_Y)$ and let $\overline{u}_i = (u_{i,1}, u_{i,2}, \ldots) \in Y$ be an approximate eigenvector for μ, i.e. $\|\overline{u}_i\| = \lim_n \|u_{i,n}\| = 1$ and

$$\lim_i \lim_n \|Su_{i,n} - \mu u_{i,n}\| = \lim_i \|\overline{S}\overline{u}_i - \mu \overline{u}_i\| = 0 \ .$$

Note that since $\overline{u}_i \in Y$ also

$$\lim_i \lim_n \|Tu_{i,n} - \lambda u_{i,n}\| = \lim_i \|\overline{T}\overline{u}_i - \lambda \overline{u}_i\| = 0 \ .$$

Finally, choose $(i_k, n_k)_{k=1}^\infty$ such that

$$\lim_k \|Tu_{i_k,n_k} - \lambda u_{i_k,n_k}\| = 0$$

and

$$\lim_k \|Su_{i_k,n_k} - \lambda u_{i_k,n_k}\| = 0$$

and put

$$u_k = u_{i_k,n_k} \ .$$

□

12.3. We turn now to the main part of this chapter.

THEOREM: *Let $\{x_i\}_{i=1}^\infty$ be an 1-unconditional 1-spreading sequence in some Banach space X. Then the unit vector basis in ℓ_p, for some $1 < p < \infty$, or in c_0 is block finitely representable on $\{x_i\}_{i=1}^\infty$.*

PROOF: For convenience of notation we change the index set from $\mathbb{N}$ to Q – the rationals in $(0,1)$: Define a norm on the space of all sequences $\{a_r\}_{r\in Q}$ with only finitely many nonzero terms by

$$\|\{a_r\}\| = \left\| \sum_{i=1}^n a_{r_i} x_i \right\| ,$$

where $r_1 < r_2 < \ldots < r_n$ contains all the indices r in Q for which $a_r \neq 0$. Denote this space by Y_0. The fact that $\{x_i\}_{i=1}^\infty$ is 1-spreading implies that the definition is consistent. Let Y be the completion of Y_0 and let $\{e_r\}_{r\in Q}$ be the unit vector basis in Y, i.e. $e_r = \{\delta_{r,s}\}_{s\in Q}$. Since $\|\Sigma_{i=1}^n a_i e_{r_i}\| = \|\sum_{i=1}^n a_i x_i\|$ for all $r_1 < \ldots < r_n$, we have that, for all $r_1 < \ldots < r_n$, $s_1 < \ldots < s_n$ in Q and signs $\varepsilon_1, \ldots, \varepsilon_n$,

$$\left\| \sum_{i=1}^n a_i e_{r_i} \right\| = \left\| \sum_{i=1}^n a_i e_{s_i} \right\|$$

and

$$\left\|\sum_{i=1}^{n} a_i e_{r_i}\right\| = \left\|\sum_{i=1}^{n} \varepsilon_i a_i e_{r_i}\right\| .$$

We are going to show that the unit vector basis of ℓ_p for some $1 \leq p < \infty$ or of c_0 is block finitely representable on $\{e_r\}_{r\in Q}$ and thus also on $\{x_i\}_{i=1}^{\infty}$.

We extend Y_0 and Y to spaces over the complex field by defining $\|\Sigma_{r\in Q} a_r e_r\| = \|\Sigma_{r\in Q} |a_r| e_r\|$ for $\{a_r\} \subseteq \mathbb{C}$. We continue to denote these spaces by Y_0 and Y.

Consider the two operators on Y_0 defined by

$$T\left(\sum_{r\in Q} a_r e_r\right) = \sum_{r\in Q} a_r e_{r/2} + \sum_{r\in Q} a_r e_{\frac{r+1}{2}}$$

and

$$S\left(\sum_{r\in Q} a_r e_r\right) = \sum_{r\in Q} a_r e_{\frac{r}{3}} + \sum_{r\in Q} a_r e_{\frac{r+1}{3}} + \sum_{r\in Q} a_r e_{\frac{r+2}{3}} .$$

Note that

$$\|x\| \leq \|Tx\| \leq 2\|x\|$$

$$\|x\| \leq \|Sx\| \leq 3\|x\| ,$$

that

$$TS = ST$$

and that T and S map nonnegative sequences to nonnegative sequences.

By Proposition 12.2, given any approximate eigenvalue λ of T there exist $\mu \in \sigma(S)$ and $\{y_n\}_{n=1}^{\infty} \in Y$, $\|y_n\| = 1$, such that $\|Ty_n - \lambda y_n\| \underset{n\to\infty}{\longrightarrow} 0$ and $\|Sy_n - \mu y_n\| \underset{n\to\infty}{\longrightarrow} 0$. The positivity of T and S implies that $|\lambda|$, $|\mu|$ and $|y_n|$ (= the sequence in Y attained from y_n by taking the absolute value of the elements of the sequence forming y_n) also satisfy the same property, so we may and shall assume that the original λ, μ and y_n are nonnegative. Necessarily $1 \leq \lambda \leq 2$, $1 \leq \mu \leq 3$. We may and shall assume also that $y_n \in Y_0$.

12.3.1. IMPORTANT REMARK: The p in the conclusion of the Theorem will depend only on the λ above; actually $2^{1/p} = \lambda$ (if $\lambda = 1$ we get c_0). This is important when we want to identify the ℓ_p (or c_0) one gets (see 12.5 and the next chapter).

We now continue with the proof. Let $u, v \in Y_0$ satisfy $max(supp\ u) < min(supp\ v)$ and let u_0^n and $u_1^n, \ldots, u_{2^k}^n$ be elements in Y_0 all with the same order distribution as y_n (i.e. if $y_n = \Sigma_{i=1}^{\ell} a_{r_i} e_{r_i}$ with $r_1 < \ldots < r_\ell$ then $u_j^n = \Sigma_{i=1}^{\ell} a_{r_i} e_{s_i}$ with $s_1 < \ldots < s_\ell$ depending on j, n) and such that

$$max(supp\ u) < min(supp\ u_i^n) < max(supp\ u_i^n) < min(supp\ v) \quad \text{for } 0 \leq i \leq 2^k$$

and

$$max(supp\ u_i^n) < min(supp\ u_{i+1}^n) \quad \text{for } 1 \leq i \leq 2^k - 1 .$$

Then

$$\left| \|u + \sum_{i=1}^{2^k} u_i^n + v\| - \|u + \lambda^k u_0^n + v\| \right| \xrightarrow[n\to\infty]{} 0 . \qquad (*)$$

This follows from the fact that $\|T^k y_n - \lambda^k y_n\| \xrightarrow[n\to\infty]{} 0$ and the invariance of the norm under spreading.

We also have a similar statement with 3^k replacing 2^k, μ replacing λ and S replacing T.

Let u_i^n be as above and define for any finite sequence $\{a_i\}_{i=1}^m$

$$|||\{a_i\}||| = \lim_n \left\| \sum a_i u_i^n \right\|$$

where the *lim* is over an appropriate subsequence (one for which the *lim* exists for all finite $\{a_i\}_{i=1}^m$).

$|||\cdot|||$ is clearly a norm. Let Z be the completion of the space of finite sequences under this norm and let $\{z_i\}_{i=1}^\infty$ be the unit vector basis. Then $\{z_i\}_{i=1}^\infty$ is block finitely representable on $\{e_r\}_{r\in Q}$. We are going to show that $\{z_i\}_{i=1}^\infty$ is 1-equivalent to the unit vector basis in either ℓ_p for some $1 \le p < \infty$ or in c_0.

First note that by $(*)$ and its analogue for 3^k we get

$$\left\| \sum_{i=1}^m a_i z_i + \sum_{i=m+1}^{m+2^k} z_i + \sum_{i=m+2^k+1}^\infty a_i z_i \right\| = \left\| \sum_{i=1}^m a_i z_i + \lambda^k z_{m+1} + \sum_{i=m+2^k+1}^\infty a_i z_i \right\| , \qquad (**)$$

$$\left\| \sum_{i=1}^m a_i z_i + \sum_{i=m+1}^{m+3^k} z_i + \sum_{i=m+3^k+1}^\infty a_i z_i \right\| = \left\| \sum_{i=1}^m a_i z_i + \mu^k z_{m+1} + \sum_{i=m+2^k+1}^\infty a_i z_i \right\| \qquad (***)$$

for all $\{a_i\}$, m and k. It is also clear that $\{z_i\}$ is 1-spreading and 1-unconditional. Equations $(**)$ and $(***)$ together imply that

$$\begin{aligned} &\left\| \sum_{i=1}^m a_i z_i + \sum_{i=m+1}^{m+2^k 3^\ell} z_i + \sum_{i=m+2^k 3^\ell+1}^\infty a_i z_i \right\| \\ &= \left\| \sum_{i=1}^m a_i z_i + \lambda^k \mu^\ell z_{m+1} + \sum_{i=m+2^k 3^\ell+1}^\infty a_i z_i \right\| \end{aligned} \qquad (****)$$

for all $\{a_i\}$, m, k and ℓ. In particular,

$$\left\| \sum_{i=1}^{2^k\cdot 3^\ell} z_i \right\| = \lambda^k \cdot \mu^\ell .$$

CASE 1: $\lambda^n = \mu^m$ for some $n, m \ge 0$ not both of which are zero. Assume, for example, $3^m > 2^n$. Then for k large enough $3^{km} > 2^{kn+1}$. It follows that

$$\lambda^{kn+1} = \left\| \sum_{i=1}^{2^{kn+1}} z_i \right\| \le \left\| \sum_{i=1}^{3^{km}} z_i \right\| = \mu^{km} = \lambda^{kn} .$$

Thus, $\lambda = 1$ and it follows that $\{z_i\}_{i=1}^{\infty}$ is isometrically equivalent to the unit vector basis of c_0.

CASE 2: $\lambda^n \neq \mu^m$ for all $n, m \in \mathbb{N}$. Then both $\lambda > 1$ and $\mu > 1$. Since $2^k \geq 3^\ell \Leftrightarrow \lambda^k \geq \mu^\ell$ we get that the function

$$f\left(\frac{2^k}{3^\ell}\right) = \frac{\lambda^k}{\mu^\ell},$$

defined for all $k, \ell \in \mathbb{N}$, is increasing. Since both $\frac{\log 2}{\log 3}$ and $\frac{\log \lambda}{\log \mu}$ are irrational (the irrationality of $\frac{\log \lambda}{\log \mu}$ follows from $\lambda^n \neq \mu^m$ for all $n, m \in \mathbb{N}$), it follows that the set of their respective integer multiplies *mod* 1 are dense in $[0,1]$ and so $\{\frac{2^k}{3^\ell}\}_{k,\ell \in \mathbb{N}}$ and $\{\frac{\lambda^k}{\mu^\ell}\}_{k,\ell \in \mathbb{N}}$ are dense in $\mathbb{R}^+$. This implies that the function f extends to a multiplicative function $\overline{f} : R^+ \to R^+$. It follows that there exists a $1 \leq p < \infty$ such that $f(x) = x^{1/p}$; in particular $\lambda = 2^{1/p}$ and $\mu = 3^{1/p}$.

Next we are going to show that

$$\left\| \sum_{i=1}^{k} z_i \right\| = k^{1/p}$$

for all k. We have already proved this for $k = 2^n \cdot 3^m$. For a general k, fix $\ell \in \mathbb{N}$ and let $n, m \in \mathbb{N}$ be such that

$$\frac{2^n}{3^m} \leq k \leq (1 + 3^{-\ell}) \cdot \frac{2^n}{3^m}. \tag{* * * * *}$$

Then,

$$2^n \cdot 3^\ell \leq k \cdot 3^{\ell+m} \leq 2^n \cdot 3^\ell + 2^n$$

and

$$2^{n/p} 3^{\ell/p} = \left\| \sum_{i=1}^{2^n 3^\ell} z_i \right\| \leq \left\| \sum_{i=1}^{k \cdot 3^{\ell+m}} z_i \right\| \leq \left\| \sum_{i=1}^{2^n \cdot 3^\ell} z_i \right\| + \left\| \sum_{i=1}^{2^n} z_i \right\| = 2^{n/p} 3^{\ell/p} + 2^{n/p}$$

or

$$2^{n/p} \cdot 3^{\ell/p} \leq 3^{\frac{\ell+m}{p}} \left\| \sum_{i=1}^{k} z_i \right\| \leq 2^{n/p} \cdot 3^{\ell/p} + 2^{n/p}.$$

Dividing by $3^{\frac{\ell+m}{p}}$ and using $(* * * * *)$ we get

$$k^{1/p}/(1 + 3^{-\ell})^{1/p} \leq (2^n/3^m)^{1/p} \leq \left\| \sum_{i=1}^{k} z_i \right\| \leq (2^n/3^m)^{1/p}(1 + 3^{-\ell/p}) \leq k^{1/p}(1 + 3^{-\ell}).$$

Sending ℓ to infinity we get

$$\left\| \sum_{i=1}^{k} z_i \right\| = k^{1/p}.$$

Finally, given a_i of the form $a_i = 2^{n_i/p} 3^{-m_i/p}$, $i = 1, \ldots, \ell$ we get from $(* * **)$ with $M = max\ m_i$

$$\left\| \sum_{i=1}^{\ell} a_i z_i \right\| = 3^{-M/p} \left\| \sum_{i=1}^{\ell} 2^{n_i/p} 3^{(M-m_i)/p} z_i \right\|$$

$$= 3^{-M/p} \left\| \sum_{i=1}^{N} z_i \right\| = 3^{-M/p} N^{1/p}$$

with $N = \Sigma_{i=1}^{\ell} 2^{n_i} 3^{M-m_i}$. So

$$\left\|\sum_{i=1}^{\ell} a_i z_i\right\| = \left(\sum_{i=1}^{\ell} 2^{n_i} 3^{-m_i}\right)^{1/p} = \left(\sum_{i=1}^{\ell} a_i^p\right)^{1/p} .$$

Since a_i's of this form are dense in the set of all coefficients the proof is complete.

□

We conclude this chapter with three corollaries. The first one is Krivine's Theorem.

12.4. THEOREM: *Let $\{x_i\}_{i=1}^{\infty}$ be a sequence in a Banach space X with the property that for all n and all scalars $a_1, \dots, a_n$, not all of which are zero, there exist constants $0 < c < C < \infty$ such that for all $i_1 < \dots < i_n$*

$$c \leq \left\|\sum_{j=1}^{n} a_j x_{i_j}\right\| \leq C .$$

Then the unit vector basis in ℓ_p for some $1 \leq p < \infty$, or in c_0 is block finitely representable on $\{x_i\}_{i=1}^{\infty}$.

PROOF: Apply Theorems 11.8 and 12.3.

□

12.5. The next proposition is a corollary to the *proof* of Theorem 12.3. This Proposition will be an important tool in the next chapter.

PROPOSITION: *Let $\{x_i\}_{i=1}^{\infty}$ be a normalized 1-unconditional 1-spreading sequence in some Banach space X with cotype q for some $q < \infty$. Assume moreover that for some $1 \leq p < \infty$, $C < \infty$ and $\varepsilon_s \searrow 0$, $s \to \infty$,*

$$s^{1/(p+\varepsilon_s)} \leq \left\|\sum_{i=1}^{s} x_i\right\| \leq C \cdot s^{1/p}$$

for all $s \geq 1$. Then the unit vector basis in ℓ_p is block finitely representable on $\{x_i\}_{i=1}^{\infty}$.

PROOF: We begin as in the proof of Theorem 12.3, defining the spaces Y_0 and Y and the operators T and S. We are going to show that, under the assumptions of the Proposition, $\lambda = 2^{1/p}$ is an approximate eigenvalue of T. Then continuing as in the proof of Theorem 12.3 (see Remark 12.3.1) we get the desired conclusion.

To show that $2^{1/p}$ is an approximate eigenvalue of T let

$$y_n = \sum_{k=1}^{2^n - 1} a_{k \cdot 2^{-n}} e_{k \cdot 2^{-n}}$$

where $a_{k\cdot 2^{-n}} = 2^{-\ell/p}$ if $k \cdot 2^{-n} = \alpha \cdot 2^{-(\ell+1)}$, α odd. That is, y_n takes the values

$$\begin{aligned}
&1 \ \text{at} \ \frac{1}{2} \\
&2^{-1/p} \ \text{at} \ \frac{1}{4} \ \text{and} \ \frac{3}{4} \\
&2^{-2/p} \ \text{at} \ \frac{1}{8}, \frac{3}{8}, \frac{5}{8} \ \text{and} \ \frac{7}{8} \\
&\vdots \\
&2^{-(n-1)/p} \ \text{at} \ \frac{1}{2^n}, \frac{3}{2^n}, \dots, \frac{2^n-1}{2^n} .
\end{aligned}$$

Now Ty_n takes the values

$$\begin{aligned}
&1 \ \text{at} \ \frac{1}{4} \ \text{and} \ \frac{3}{4} \\
&2^{-1/p} \ \text{at} \ \frac{1}{8}, \frac{3}{8}, \frac{5}{8} \ \text{and} \ \frac{7}{8} \\
&\vdots \\
&2^{-(n-2)p} \ \text{at} \ \frac{1}{2^n}, \frac{3}{2^n}, \dots, \frac{2^n-1}{2^n}
\end{aligned}$$

and

$$2^{-(n-1)/p} \ \text{at} \ \frac{1}{2^{n+1}}, \frac{3}{2^{n+1}}, \dots, \frac{2^{n+1}-1}{2^{n+1}} .$$

Consequently,

$$Ty_n - 2^{1/p}y_n = 2^{-(n-1)/p}\left(e_{2^{-(n+1)}} + e_{3\cdot 2^{-(n+1)}} + \dots + e_{(2^{n+1}-1)2^{-(n+1)}}\right) - 2^{1/p}e_{1/2}$$

so that

$$\|Ty_n - 2^{1/p}y_n\| \le C \cdot 2^{1/p} + 2^{1/p} = (C+1)2^{1/p} .$$

Now, using the cotype q inequality, we get

$$\begin{aligned}
\|y_n\| &\ge C_q^{-1}(X)\left(\sum_{j=0}^{n-1}\left(2^{-j/p}\left\|\sum_{i=1}^{2^j} x_i\right\|\right)^q\right)^{1/q} \\
&\ge C_q^{-1}(X)\left(\sum_{j=0}^{n-1}\left(2^{-j/p}2^{j/(p+\epsilon_{2^j})}\right)^q\right)^{1/q} .
\end{aligned}$$

Denote $\delta_n = \frac{1}{p} - \frac{1}{p+\epsilon_{2^{n-1}}}$. Then $\delta_n \to 0$ and

$$\|y_n\| \ge C_q^{-1}(X) \cdot \left(\sum_{j=0}^{n-1} 2^{-\delta_n \cdot j \cdot q}\right)^{1/q} \underset{n\to\infty}{\longrightarrow} \infty$$

(if $\alpha_n \to 1$ then $1 + \alpha_n + \alpha_n^2 + \dots + \alpha_n^{n-1} \underset{n\to\infty}{\longrightarrow} \infty$). Finally, put $z_n = y_n/\|y_n\|$, then $\|z_n\| = 1$ and

$$\|Tz_n - 2^{1/p} z_n\| \le (C+1)2^{1/p}/\|y_n\| \underset{n\to\infty}{\longrightarrow} 0 \ .$$

□

12.6. The last proposition is a simple consequence of Proposition 12.5. We state it in the form to be used in the next chapter.

PROPOSITION: *Let $\delta_n \searrow 0$ and let $\{B_n\}_{n=1}^{\infty}$ be a sequence of finite dimensional normed spaces with a common estimate on their cotype q constant for some $q < \infty$. Assume that B_n is spanned by a sequence $y_1^n, \ldots, y_n^n$ of norm one elements which is $(1+\delta_n)$-spreading and C-unconditional for some C independent of n. If for every sequence $t_n \to \infty$, $t_n \le n$ of integers*

$$p = inf \left\{ r;\ \lim_{n\to\infty} t_n^{-1/r} \left\| \sum_{i=1}^{t_n} y_i^n \right\| = \infty \right\}$$

then ℓ_p is block finitely representable on $\{B_n\}_{n=1}^{\infty}$, i.e., for every $\varepsilon > 0$ and k there exists an N such that for $n > N$ there are k blocks of $\{y_1^n, \ldots, y_n^n\}$ which are $(1+\varepsilon)$-equivalent to the unit vector basis of ℓ_p^k.

13. THE MAUREY-PISIER THEOREM

13.1. Given an infinite dimensional Banach space X and $1 \le p \le 2 \le q < \infty$, recall that $T_p(X)$ (resp. $C_q(X)$) denotes the type p (resp. cotype q) constant of X (see 9.1).

Let

$$p_X = sup\{p; T_p(X) < \infty\}$$

$$q_X = inf\{q; C_q(X) < \infty\} .$$

Note, the sup and inf need not be attained so that X is not necessarily of type p_X or of cotype q_X. Most of this chapter is devoted to the proof of the following theorem of Maurey and Pisier. This is a stronger version of Krivine's Theorem (12.4).

13.2. THEOREM: *Let X be an infinite dimensional Banach space. Then ℓ_{p_X} and ℓ_{q_X} are finitely representable in X.*

We give a detailed proof of the cotype case (i.e. that ℓ_{q_X} is finitely representable in X). A similar but somewhat more complicated proof can be given for the type case (see [M.S.]). We preferred, however, to sketch a different proof for this case which has the advantage of giving a reasonable estimation on the dimension of the embedded ℓ_p^k space in the finite dimensional version of the theorem (see 13.12 below), but has the disadvantage of not working in the cotype case.

13.3. We begin with the proof for the cotype case. For a fixed Banach space X we shall denote $T_p(n) = T_p(X,n), C_q(n) = C_q(X,n)$, i.e., $T_p(n)$ is the smallest T for which

$$\left(\int_0^1 \left\|\sum_{i=1}^n r_i(t)x_i\right\|^2 dt\right)^{1/2} \le T\left(\sum_{i=1}^n \|x_i\|^p\right)^{1/p}$$

for all $x_1,\ldots,x_n \in X$. Similarly $C_q(n)$ is the smallest C for which

$$\left(\sum_{i=1}^n \|x_i\|^q\right)^{1/q} \le C\left(\int_0^1 \left\|\sum_{i=1}^n r_i(t)x_i\right\|^2\right)^{1/2}$$

for all $x_1,\ldots,x_n$.

13.4. LEMMA: *For each $1 \le p \le 2$ and $2 \le q < \infty$ the sequences $\{C_q(n)\}_{n=1}^\infty$ and $\{T_p(n)\}_{n=1}^\infty$ are sub- multiplicative, i.e. for all integers n,k we have*

(a) $$C_q(nk) \le C_q(n)C_q(k)$$

(b) $$T_p(nk) \le T_p(n)T_p(k) .$$

PROOF: (a) Let $\{x_i\}_{i=1}^{nk}$ be nk elements of X. For each $1 \leq i \leq n$ and $\theta \in [0,1]$, let us define $X_i(\theta) = \sum_{(i-1)k<j\leq ik} r_j(\theta)x_j$. From the definition of $C_q(n)$ we have

$$\sum_{i=1}^{n} \|X_i(\theta)\|^q \leq (C_q(n))^q \int \left\|\sum_{i=1}^{n} r_i(t)X_i(\theta)\right\|^q dt.$$

Integrating with respect to θ, we obtain

$$\begin{aligned}\sum_{i=1}^{n} \int \|X_i(\theta)\|^q d\theta &\leq (C_q(n))^q \int \left\|\sum_{i=1}^{n} r_i(t)X_i(\theta)\right\|^q d\theta dt \\ &= (C_q(n))^q \int \left\|\sum_{j=1}^{nk} r_j(\theta)x_j\right\|^q d\theta\end{aligned}$$

which follows easily from the symmetry of the Rademacher variables. On the other hand, for each $1 \leq i \leq n$,

$$\sum_{(i-1)k<j\leq ik} \|x_j\|^q \leq (C_q(k))^q \int \|X_i(\theta)\|^q d\theta$$

and the assertion follows immediately from the last two inequalities.

(b) The proof is completely analogous to the proof of (a).

□

13.5. LEMMA: *If $N > 1$ is an integer, and $q \geq r \geq 2$ is defined by $C_r(N) = N^{1/r-1/q}$, then X is of cotype s for every $s > q$, i.e. $q_X \leq q$.*

(Similarly, if $N > 1$ and $p \leq r \leq 2$ is defined by $T_r(N) = N^{1/p-1/r}$, then X is of type s, for each $s < p$; we will not prove this part since, as we said before, our proof of Theorem 13.2 for the type case will follow a different route).

PROOF: From the monotonicity and submultiplicativity of the sequence $\{C_r(n)\}_{n=1}^{\infty}$, it follows that there exists a constant $C > 0$ such that, for all integers n,

$$C_r(n) \leq C \cdot n^{\frac{1}{r}-\frac{1}{q}}.$$

Now, let $\{x_i\}_{i=n}^{n}$ be n elements of X, and with no loss of generality, assume that their norms are in decreasing order. Then, for each $1 \leq k \leq n$,

$$\begin{aligned}\|x_k\| = \inf_{j\leq k} \|x_j\| \leq \left(\frac{1}{k}\sum_{j=1}^{k} \|x_j\|^r\right)^{\frac{1}{r}} &\leq k^{-\frac{1}{r}} C_r(k) \left(\int \left\|\sum_{j=1}^{k} r_j(t)x_j\right\|^r dt\right)^{\frac{1}{r}} \\ &\leq C \cdot k^{-\frac{1}{q}} \left(\int \left\|\sum_{j=1}^{n} r_j(t)x_j\right\|^r dt\right)^{\frac{1}{r}}\end{aligned}$$

which follows from the triangle inequality in $L^q(X)$. Thus, for each $s > q$,

$$\left(\sum_{k=1}^{n} \|x_k\|^s\right)^{\frac{1}{s}} \leq C \left(\sum_{k=1}^{\infty} k^{\frac{-s}{q}}\right)^{\frac{1}{s}} \cdot \left(\int \left\|\sum_{j=1}^{n} r_j(t)x_j\right\|^r dt\right)^{\frac{1}{r}}$$

$$\leq C \cdot C_1(s,q,r) \left(\int \left\|\sum_{j=1}^{n} r_j(t)x_j\right\|^s dt\right)^{\frac{1}{s}}$$

by Kahane's Theorem 9.2. So X is of cotype s.

$\square$

13.6. COROLLARY: *For each integer n and p, q with $p_X \leq p \leq 2 \leq q \leq q_X$,*

$$C_q(n) \geq n^{\frac{1}{q} - \frac{1}{q_X}}$$

$$\left(and \ T_p(n) \geq n^{\frac{1}{q_X} - \frac{1}{p}}\right).$$

13.7. LEMMA: If $q \leq q_X$, then $\lim_{n \to \infty} \frac{\log C_q(n)}{\log n} = \frac{1}{q} - \frac{1}{q_X}$.
(Similarly, and again without proof, if $p \geq p_X$, then $\lim_{n \to \infty} \frac{\log T_p(n)}{\log n} = \frac{1}{p_X} - \frac{1}{p}$).

PROOF: It follows from the last corollary that

$$\lim_{n \to \infty} \frac{\log C_q(n)}{\log n} \geq \frac{1}{q} - \frac{1}{q_X}.$$

On the other hand, let $\varepsilon > 0$, $r > q_X$, and let n be an integer. By the definition of $C_q(n)$ there exist $y_1, \ldots, y_n$ such that

$$(1 - \varepsilon)C_q(n) \left(\int \left\|\sum_{i=1}^{n} y_i r_i(t)\right\|^q dt\right)^{\frac{1}{q}} \leq \left(\sum_{i=1}^{n} \|y_i\|^q\right)^{\frac{1}{q}} \leq n^{\frac{1}{q} - \frac{1}{r}} \left(\sum_{i=1}^{n} \|y_i\|^r\right)^{\frac{1}{r}}$$

$$\leq C_r n^{\frac{1}{q} - \frac{1}{r}} \left(\int \left\|\sum_{i=1}^{n} y_i r_i(t)\right\|^r dt\right)^{\frac{1}{r}} \leq C_r' n^{\frac{1}{q} - \frac{1}{r}} \left(\int \left\|\sum_{i=1}^{n} y_i r_i(t)\right\|^q dt\right)^{\frac{1}{q}}$$

for some constant C_r' (since X is of cotype r and by Kahane's inequality 9.2). Hence, $C_q(n) \leq \frac{1}{1-\varepsilon} C_r n^{1/q - 1/r}$ and, therefore,

$$\overline{\lim}_{n \to \infty} \frac{\log C_q(n)}{\log n} \leq \frac{1}{q} - \frac{1}{r}.$$

Since $r > q_X$ was arbitrary, the proof is complete.

$\square$

13.8. We now begin the proof of the cotype case of Theorem 13.2. If $q_X = 2$, then, by Dvoretzky's Theorem 5.8, we are done. Assume $2 < q_X \leq \infty$ (the case $q_X = \infty$ is much simpler but we prefer to treat it in the general scheme). It suffices to prove the following claim.

CLAIM: *There exist constants $0 < \gamma, C < \infty$ and $0 < \delta < 1$ such that for each $\varepsilon > 0$ and integer m there exist m elements $y_1, \ldots, y_m$ of X with*

(i) $$1 - \delta \le \|y_i\| \le 1, \quad i = 1, \ldots, m$$

(ii) $(y_1, \ldots, y_m)$ *is an unconditional basic sequence with constant $\le C$ and is $(1+\varepsilon)$-invariant to spreading.*

(iii) $$\|\Sigma_{j=1}^{s} y_j\| \le \gamma \cdot s^{\frac{1}{q_X}}, \text{ for all } s \le m.$$

Indeed, suppose that the claim is true. In the case $q_X = \infty$ the inequality *(iii)* implies $\|\Sigma_1^m a_j y_j\| \le \gamma \max_{1 \le j \le m} |a_j|$ and $span\{y_j\}_{j=1}^m$ is $(C \cdot \gamma)$-isomorphic to ℓ_∞^m. Use now the lemma in 11.8 to finish the proof in this case. If $q_X < \infty$, then we are in a position to apply Krivine's theorem. For each k, let $y_1^k, \ldots, y_k^k$ be k elements of X satisfying the claim with $\varepsilon_k = \frac{1}{k}$, say. Define $z_j^k = y_j^k / \|y_j^k\|$, $j = 1, \ldots, k$. Then, for any $q_k > q_X$, we shall have, for every $s \le k$,

$$s^{\frac{1}{q_k}} = \left(\sum_{j=1}^{s} \|z_j^k\|^{q_k} \right)^{\frac{1}{q_k}} \le C_{q_k}(X) \left(\int \left\| \sum_{j=1}^{s} z_j^k r_j(t) \right\|^{q_k} dt \right)^{\frac{1}{q_k}}$$

$$\le C_{q_k}(X) C \left\| \sum_{j=1}^{s} z_j^k \right\| \le \frac{\gamma \cdot C}{1 - \delta} C_{q_k}(X) s^{\frac{1}{q_X}} .$$

Thus, for any r, and each $s \le k$,

$$\frac{s^{\frac{1}{q_k} - \frac{1}{r}}}{C \cdot C_{q_k}(X)} \le s^{-\frac{1}{r}} \left\| \sum_{j=1}^{s} z_j^k \right\| \le \frac{\gamma}{1 - \delta} s^{\frac{1}{q_X} - \frac{1}{r}} .$$

Now it follows that if $r < q_X$ then, for every sequence of integers $k \ge s_k \to \infty$,

$$\lim_k s_k^{-\frac{1}{r}} \left\| \sum_{j=1}^{s_k} z_j^k \right\| = 0 .$$

On the other hand one can always find a sequence $\{q_k\}_{k=1}^\infty$ such that $q_k > q_X$ for all k, $\lim_k q_k = q_X$ and

$$\lim_k \frac{s_k^{\frac{1}{q_X} - \frac{1}{q_k}}}{C_{q_k}(X)} = \infty.$$

Let $r > q_X$ then, if k is large enough, $s_k^{\frac{1}{q_k} - \frac{1}{r}} \ge s_k^{\frac{1}{q_X} - \frac{1}{q_k}}$ and

$$\lim_k s_k^{-\frac{1}{r}} \left\| \sum_{j=1}^{s_k} z_j^k \right\| = \infty.$$

Hence, ℓ_{q_X} is indeed finitely representable in X, by Proposition 12.6.

13.9. PROOF OF THE CLAIM: Assume first $q_X < \infty$. Let k be given, and let us fix $2 \leq r = q_X - \frac{h}{\ell n k} < q$, where $h > 0$ is chosen so that $k^{\frac{1}{r} - \frac{1}{q_X}} \leq 2$. Put $\alpha = \frac{1}{2}(1 - \frac{r}{q_X}) = \frac{h}{2 q_X \ell n k}$, and choose n large enough, so that $n^\alpha \geq N(k, \varepsilon)$ of Proposition 11.9. In the case $q_X = \infty$ we choose r so that $k^{1/r} \leq 2$ and take $\alpha = 1/2$. Then, by the definition of $C_r(n)$, there exist $x_1, \ldots, x_n \in X$ such that

13.9.1
$$\left(\sum_{i=1}^{n} \|x_i\|^r\right)^{1/r} > (1-\varepsilon) C_r(n) \left(\int \|\sum_{i=1}^{n} x_i r_i(t)\|^r dr\right)^{1/r}.$$

With no loss of generality, assume that $\max_i \|x_i\| = 1$. Let $0 < \delta < 1$ and put $w = 1 - \delta$. Define, for each $j \geq 1$ and $i \geq 1$,

$$A_j = \{i : w^j < \|x_i\| \leq w^{j-1}\}$$

$$\hat{x}_i = x_i / w^{j-1}, \qquad for\ i \in A_j.$$

If $|A_j| \leq n^\alpha$, we set $A_{j,0} = A_j$ and do nothing else with this set. If $|A_j| > n^\alpha$, then we choose a subset $A_{j,1} \subset A_j$ for which

$$\left(\int \left\|\sum_{i \in A_{j,1}} \hat{x}_i r_i(t)\right\|^r dt\right)^{\frac{1}{r}} \Big/ |A_{j,1}|^{\frac{1}{q_X}} = \max_{T \subseteq A_j} \left(\int \left\|\sum_{i \in T} \hat{x}_i r_i(t)\right\|^r\right)^{\frac{1}{r}} \Big/ |T|^{\frac{1}{q_X}}.$$

If $|A_j - A_{j,1}| > n^\alpha$, then there exists another subset $A_{j,2} \subset A_{j,2}^{(1)} = A_j - A_{j,1}$ with the same properties. Continuing in this manner we obtain a disjoint partition $A_j = A_{j,1} \cup \ldots \cup A_{j,m_j} \cup A_{j,0}$, where $|A_{j,0}| \leq n^\alpha$ and for each $1 \leq p \leq m_j$, $A_{j,p} \subset A_{j,p}^{(1)}$, where $A_{j,p}^{(1)}$ contains more than n^α elements, and for each $T \subset A_{j,p}^{(1)}$,

$$\frac{\left(\int \left\|\sum_{i \in A_{j,p}} \hat{x}_i r_i(t)\right\|^r dt\right)^{\frac{1}{r}}}{|A_{j,p}|^{\frac{1}{q_X}}} \geq \frac{\left(\int \left\|\sum_{i \in T} \hat{x}_i r_i(t)\right\|^r dt\right)^{\frac{1}{r}}}{|T|^{\frac{1}{q_X}}}$$

We now split $\{1, \ldots, n\}$ into two parts. The "bad" part is $A' = \cup_j A_{j,0}$ and the "good" part, its complement, is $A'' = \cup_j \cup_{1 \leq t \leq m_j} A_{j,t}$. Let us show first that the "bad" part can be essentially ignored in inequality 13.9.1. Indeed,

$$\begin{aligned}\sum_{i=1}^{n} \|x_i\|^r &= \sum_{i \in A'} \|x_i\|^r + \sum_{i \in A''} \|x_i\|^r \\ &\leq \sum_j w^{(j-1)r} |A_{j,0}| + \sum_{i \in A''} \|x_i\|^r \leq \frac{n^\alpha}{1 - w^r} + \sum_{i \in A''} \|x_i\|^r.\end{aligned}$$

But α is chosen so that, by the property of $C_r(n)$ in Corollary 13.6, $n^\alpha = o(C_r(n)^r)$, and since $1 \leq \int \|\sum_{i=1}^n x_i r_i(t)\|^r dt$ (because $max\|x_i\| = 1$, and the triangle inequality), we may assume that

$$\frac{n^\alpha}{1 - w^r} < \varepsilon(1-\varepsilon)^r C_r(n)^r \int \left\|\sum_{i=1}^{n} x_i r_i(t)\right\|^r dt.$$

Thus, it follows from 13.9.1 that

$$(1-\varepsilon)^{r+1}C_r(n)^r \int \left\| \sum_{i=1}^n x_i r_i(t) \right\|^r dt \le \sum_{i\in A''} \|x_i\|^r,$$

and by the symmetry properties of the Rademacher variables,

$$(1-\varepsilon)^{r+1}C_r(n)^r \int \left\| \sum_{i\in A''} x_i r_i(t) \right\|^r dt \le \sum_{i\in A''} \|x_i\|^r$$

so that we have a similar inequality to 13.9.1, involving only the "good" part of $\{1,\ldots,n\}$.

Let us now introduce new symbols $\{B_s\}_{s=1}^{\ell}$ for the subsets $\{A_{j,t}\}_{j,1\le t\le m_j}$. Let us define

$$u_s(\theta) = \sum_{i\in B_s} x_i r_i(\theta),\ s\le \ell,\ \theta\in[0,1].$$

Then

$$\sum_{s=1}^{\ell} \|u_s(\theta)\|^r \le C_r(\ell)^r \int \left\| \sum_{s=1}^{\ell} \sum_{i\in B_s} x_i r_i(\theta) r_s(t) \right\|^r dt .$$

Integrating with respect to θ, and using the symmetry of the Rademacher variables, we obtain

$$\sum_{s=1}^{\ell} \int \|u_s(\theta)\|^r d(\theta) \le C_r(\ell)^r \int \left\| \sum_{i\in A''} x_i r_i(\theta) \right\|^r d\theta$$

$$\le \frac{C_r(\ell)^r}{(1-\varepsilon)^{r+1}C_r(n)^r} \sum_{i\in A''} \|x_i\|^r = \frac{C_r(\ell)^r}{(1-\varepsilon)^{r+1}C_r(n)^r} \sum_{s=1}^{\ell} \left(\sum_{i\in B_s} \|x_i\|^r \right) .$$

Hence, there must exist at least one index $i \le s_0 \le \ell$ such that

$$\int \|u_{s_0}(\theta)\|^r d\theta = \int \left\| \sum_{i\in B_{s_0}} x_i r_i(\theta) \right\|^r d\theta \le \frac{C_r(\ell)^r}{(1-\varepsilon)^{r+1}C_r(n)^r} \sum_{i\in B_{s_0}} \|x_i\|^r .$$

Replacing the x_i's by the normalized $\hat{x}_i$'s, and taking ε small enough, we have (observing that $C_r(\ell) \le C_r(n)$):

$$\left(\int \left\| \sum_{i\in B_{s_0}} \hat{x}_i r_i(\theta) \right\|^r d(\theta) \right)^{\frac{1}{r}} \le 2 \left(\sum_{i\in B_{s_0}} \|\hat{x}_i\|^r \right)^{\frac{1}{r}} \le 2|B_{s_0}|^{\frac{1}{r}}.$$

But $|B_{s_0}| \le k$, and by our choice of the number r, we obtain

$$\left(\int \left\| \sum_{i\in B_{s_0}} \hat{x}_i r_i(\theta) \right\|^r d\theta \right)^{\frac{1}{r}} \le 4|B_{s_0}|^{\frac{1}{q\chi}}$$

(in the case $q_X = \infty$ we have formally the same inequality because $|B_{s_0}|^{\frac{1}{r}} \le k^{\frac{1}{r}} \le 2$). Since the set B_{s_0} is one of the sets $A_{j,t}$ it is contained in some set $\tilde{A} \equiv A^{(1)}_{j_0,t_0}$ which consists of at least $N(k,\varepsilon)$ elements, where $N(k,\varepsilon)$ is as in Proposition 11.9. By the choice of B_{s_0}, we also obtain, for each $T \subset \tilde{A}$,

13.9.2
$$\left(\int \left\|\sum_{i\in T} \hat{x}_i r_i(\theta)\right\|^r\right)^{\frac{1}{r}} \le 4|T|^{\frac{1}{q_X}} .$$

Now use Proposition 11.9. There exists a subset $A \subset \tilde{A}$ of k elements which satisfies one of the two alternatives there. However 13.9.2 eliminates the second one for k large enough. Therefore we obtain a set A of k elements that are $(1+\varepsilon)$ -invariant to spreading and, for each $T \subset A$, satisfies 13.9.2. To complete the proof of the claim, it remains to show the existence of a set $T \subset A$, which is large enough and forms a "good" unconditional basic sequence. To do that we make use of Proposition 11.10. Let $T_0 \subset A$ be a subset such that $\{\hat{x}_i\}_{i\in T_0}$ is contained in a ball of radius δ. Since they all have norm $\ge 1-\delta$, it follows that there exists a functional $x^* \in X^*$ such that $\|x^*\| = 1$ and $|x^*(\hat{x}_i)| \ge 1 - 3\delta$ for each $i \in T_0$. Then, from 13.9.2 we obtain

$$\left(\int \left\|\sum_{i\in T_0} x^*(\hat{x}_i) r_i(\theta)\right\|^r d\theta\right)^{\frac{1}{r}} \le \left(\int \left\|\sum_{i\in T_0} \hat{x}_i r_i(\theta)\right\|^r d\theta\right) \le 4|T_0|^{\frac{1}{q_X}} .$$

On the other hand for any scalars $\{\alpha_i\}$ and any $r \ge 2$,

$$\left(\int \left|\sum_i \alpha_i r_i(\theta)\right|^r d\theta\right) \ge \left(\sum_i |\alpha_i|^2\right)^{\frac{1}{2}} .$$

Hence, $(1-3\delta)|T_0|^{\frac{1}{2}} \le 4|T_0|^{\frac{1}{q_X}}$ and so $|T_0| \le \left(\frac{4}{1-3\delta}\right)^{\frac{2q_X}{q_X-2}}$. Fix any $\delta < \frac{1}{3}$ (say $\delta = \frac{1}{6}$) and put $M = 8^{2q_X/(q_X-2)}$. To conclude, no more than a fixed number M of elements of $\{\hat{x}_i\}_{i\in A}$ can be contained in a ball of radius $\frac{1}{6}$. Hence, by a standard argument, there is a subset $T \subset A$ of at least $\frac{k}{M}$ points such that for any $i \ne j$ in T, $\|\hat{x}_i - \hat{x}_j\| \ge \frac{1}{6}$. By proposition 11.10, if k is large enough, we can find a subsequence $\hat{x}_{i_1}, \ldots, \hat{x}_{i_{2m}}$ such that the sequence $v_j \equiv \hat{x}_{i_{2j}} - \hat{x}_{i_{2j-1}}$, $j = 1,\ldots,m$, is unconditional with a constant ≤ 3, say, and $\|v_j\| \ge \frac{1}{6}$ for all $j \le m$. By 13.9.2, we obtain, for each $s \le m$,

$$\begin{aligned}\left\|\sum_{j=1}^s v_j\right\| &\le 3\left(\int \left\|\sum_{j=1}^s v_j r_j(\theta)\right\|^r d\theta\right)^{\frac{1}{r}} \\ &\le 3\left(\int \left\|\sum_{j=1}^s \hat{x}_{i_{2j}} r_j(\theta)\right\|^r d\theta\right)^{\frac{1}{r}} + 3\left(\int \left\|\sum_{j=1}^s \hat{x}_{i_{2j-1}} r_j(\theta)\right\|^r d\theta\right)^{\frac{1}{r}} \\ &\le 24 s^{\frac{1}{q_X}} ,\end{aligned}$$

and the claim follows, by setting $y_j = \frac{1}{2}v_j$, $j \le m$, (with $\delta = \frac{11}{12}$, $\gamma = 12$, and $C = 3$).

□

13.10. We now turn over to the proof of the type case of Theorem 13.2. As we mentioned above, rather than bringing a proof similar to the cotype case we prefer to briefly sketch a completely different approach due to Pisier [Pi2].

DEFINITION: For $1 < p \le 2$, the *stable type p constant* of a Banach space X, denoted by $ST_p(X)$, is the smallest constant C such that

$$E\left\|\sum \theta_i x_i\right\| \le C\left(\sum \|x_i\|^p\right)^{\frac{1}{p}}$$

for all finite sequences $\{x_i\}$ in X, where $\{\theta_i\}_{i=1}^{\infty}$ is a sequence of independent symmetric p-stable random variables normalized in L_1. For $p = 1$ one defines the stable type 1 constant of X in a similar way replacing $E\|\Sigma\theta_i x_i\|$ (which is ∞ unless $x_i = 0$ for all i) with $(E\|\Sigma\theta_i x_i\|^{\frac{1}{2}})^2$.

It is easily checked that $ST_p(X)$ is also equal to the smallest constant D such that

13.10.1
$$E\left\|\sum_{i=1}^{n} \theta_i x_i\right\| \le D \cdot \max_{1\le i\le n} \|x_i\| \cdot n^{\frac{1}{p}}$$

for all n and all $x_1, \dots, x_n \in X$. Indeed if 13.10.1 holds and $k_1, \dots, k_n$ are any positive integers then for any $x_1, \dots, x_n$, norm one elements in X, we have ($(\theta_{i,j})$ is a double sequence of normalized symmetric p-stable random variables).

$$E\left\|\sum_{i=1}^{n} \theta_i k_i^{\frac{1}{p}} x_i\right\| = E\left\|\sum_{i=1}^{n}\sum_{j=1}^{k_i} \theta_{i,j} x_i\right\| \le$$

$$\le D\left(\sum_{i=1}^{n} k_i\right)^{\frac{1}{p}}.$$

Thus, if $x_1, \dots, x_n$ are in X with norms $\|x_i\| = (k_i/s)^{\frac{1}{p}}$ for some integers k_i's then

$$E\left\|\sum \theta_i x_i\right\| \le D\left(\sum \|x_i\|^p\right)^{\frac{1}{p}}.$$

It follows now that this inequality holds for any x_i's, i.e. $ST_p(X) \le D$. The other side inequality is trivial.

13.11. Let $\{r_i\}_{i=1}^{\infty}$ be a Rademacher sequence independent of $\{\theta_i\}_{i=1}^{\infty}$. Then the sequence $\{|\theta_i| r_i\}_{i=1}^{\infty}$ has the same distribution as the sequence $\{\theta_i\}_{i=1}^{\infty}$ and by the triangle inequality

$$\int_0^1 \left\|\sum r_i(t) x_i\right\| = \int_0^1 \left\|E\sum r_i(t)|\theta_i| x_i\right\|$$
$$\le \int_0^1 E\left\|\sum r_i(t)|\theta_i| x_i\right\| = E\left\|\sum \theta_i x_i\right\| .$$

We conclude that, for $1 < p \le 2$,

13.11.1. $$T_p(X) \le K \cdot ST_p(X) \text{ for some universal } K.$$

On the other hand fix $1 \le r < p \le 2$ and let $\{\varphi_i\}$ be a sequence of independent symmetric r-stable random variables normalized in L_1 and independent of $\{r_i\}$ and $\{\theta_i\}$. Then for any $\{x_i\} \subseteq X$

$$\begin{aligned} E\left\|\sum \varphi_i x_i\right\| &= \int_0^1 E\left\|\sum |\varphi_i| r_i(t) x_i\right\| \le T_p(X) E\left(\sum |\varphi_i|^p \|x_i\|^p\right)^{\frac{1}{p}} \\ &= T_p(X) E\left|\sum \varphi_i \theta_i \|x_i\|\right| = T_p(X)\left(\sum |\theta_i|^r \|x_i\|^r\right)^{\frac{1}{r}} \\ &\le T_p(X)(E|\theta_1|^r)^{\frac{1}{r}}\left(\sum \|x_i\|^r\right)^{\frac{1}{r}} \end{aligned}$$

so that

13.11.2. $$ST_r(X) \le T_p(X)(E|\theta_1|^r)^{\frac{1}{r}}.$$

We conclude that, if $p_X < 2$ then

13.11.3. $$p_X = inf\ \{1 < p < 2,\ ST_p(X) = \infty\}.$$

13.12. THEOREM: *For any $1 < p < 2$ and $\varepsilon > 0$ there exists a $\delta = \delta(\varepsilon, p)$ such that if*

$$k < \delta(ST_p(X))^q \qquad \left(\frac{1}{p} + \frac{1}{q} = 1\right)$$

then X contains a $(1+\varepsilon)$-isomorphic copy of ℓ_p^k.

Theorem 13.12 implies the type case of Theorem 13.2. Indeed, if $p_X = 2$ we are done by Dvoretzky's Theorem. Otherwise $ST_p(X) = \infty$ for any $p > p_X$. So that X contains $(1+\varepsilon)$-isomorphic copies of ℓ_p^k for any k, ε and $p > p_X$. It follows that ℓ_{p_X} is finitely representable in X.

It is quite easy to show that $ST_p(\ell_1^n) = n^{\frac{1}{q}}$ so that another corollary to Theorem 13.12 is Theorem 8.8.

To prove Theorem 13.12 note first that we may assume $ST_p(X) < \infty$ (if $ST_p(X) = \infty$ X contains subspaces with arbitrarily large finite stable type p constant). Let $x_1, \dots, x_n \in X$ be such that $max_{1 \le i \le n}\|x_i\| \le 1$ and

$$E\left\|\sum_{i=1}^n \theta_i x_i\right\| \ge \frac{n^{\frac{1}{p}}}{2} ST_p(X).$$

Let Y be the random variable which takes each of the $2n$ (vector) values, $x_1, \dots, x_n$, $-x_1, \dots, -x_n$ with probability $\frac{1}{2n}$ and let $\{Y_{i,j}\}_{i=1}^{\infty}{}_{j=1}^{\infty}$ be a double sequence of independent (X valued) randon variables all with the same distribution as Y. For a δ to be chosen later, for $k \le \delta(ST_p(X))^q$ and for any $a = (a_1, \dots, a_k)$ with $\Sigma_{i=1}^k |a_i|^p = 1$ we form

$$S_a = \sum_{i=1}^k a_i \sum_{j=1}^{\infty} j^{-\frac{1}{p}} Y_{i,j}.$$

The proof of Theorem 13.12 is a standard consequence (see e.g. 4.1 and 8.8) of the following two propositions.

13.13. PROPOSITION: *For an appropriately chosen $\delta = \delta(\varepsilon, p)$,*

(*a*) $$E\|S_a\| \leq (1+\varepsilon)E\|S_b\|$$

for all a, b in the sphere of ℓ_p^k.

(*b*) $$E\|S_a\| \geq c_p \cdot ST_p(X)$$

for all a in the sphere of ℓ_p^k, where $c_p > 0$ depends on p only.

13.14. PROPOSITION: *For an appropriately chosen $\delta = \delta(\varepsilon, p)$, and for any a in the sphere of ℓ_p^k*

$$1 - P\left((1-\varepsilon)E\|S_a\| \leq \|S_a\| \leq (1+\varepsilon)E\|S_a\|\right) \leq 2\, exp(-\delta(ST_p(X))^q).$$

13.15. For the proofs of the two propositions we need a representation theorem for p-stable random variables from [L.W.Z.]. We refer the reader to [Mar.P.] for a proof.

Let $\{A_i\}_{i=1}^{\infty}$ be independent copies of an exponential random variable, i.e.,

$$P(A_i > t) = e^{-t}, \quad t > 0, \quad i = 1, 2, \ldots$$

and let

$$\Gamma_j = \sum_{i=1}^{j} A_i.$$

PROPOSITION: *For each $1 < p < 2$ there exists a constant $\infty > c_p > 0$ such that, if y is any symmetric random variable with $E|y|^p = 1$ and $\{y_i\}$ are independent copies of y which are independent of the A_i's, then*

$$X = \sum_{j=1}^{\infty} \Gamma_j^{-\frac{1}{p}} y_j$$

is a symmetric p-stable random variable with $E|X| = c_p$.

Let $\{\Gamma_{i,j}\}_{j=1}^{\infty}$, $i = 1, 2, \ldots$, be independent copies of $\{\Gamma_j\}_{j=1}^{\infty}$ which are also independent of $\{Y_i\}_{i=1,\ j=1}^{\infty\ \ \infty}$. Form

$$\tilde{S}_i = \sum_{j=1}^{\infty} \Gamma_{i,j}^{-\frac{1}{p}} Y_{i,j}\ ,$$

and for $a = (a_1, \ldots, a_k)$ with $\Sigma_{i=1}^k |a_i|^p = 1$ form

$$\tilde{S}_a = \sum_{i=1}^{k} a_i \tilde{S}_i.$$

For any $f \in X^*$, $(E|f(Y)|^p)^{\frac{1}{p}} = (\frac{1}{n}\Sigma_{i=1}^n |f(x_i)|^p)^{\frac{1}{p}}$.

It follows from the proposition above that $f(\tilde{S}_i)$ is a symmetric p-stable with $E|f(\tilde{S}_i)| = c_p \cdot (\frac{1}{n}\Sigma_{i=1}^n |f(x_i)|^p)^{\frac{1}{p}}$ and thus also $f(\tilde{S}_a)$ is a symmetric p-stable with the same expectation - $c_p \cdot (\frac{1}{n}\Sigma_{i=1}^n |f(x_i)|^p)^{\frac{1}{p}}$. The variable $f(\Sigma_{i=1}^n \theta_i x_i)$ is also a symmetric p-stable random variable. Its expectation is $E|f(\Sigma_{1=1}^n \theta_i x_i)| = (\Sigma |x_i|^p)^{\frac{1}{p}}$. It follows that the two processes

$$\{f(\tilde{S}_a)\}_{f\in X^*} \; and \; \{c_p \cdot f(\frac{1}{n}\sum_{i=1}^n \theta_i x_i)\}_{f\in X^*}$$

have the same distribution (compare distributions of identical linear combinations of the two processes). In particular for each a in the sphere of ℓ_p^k.

13.15.1
$$\|\tilde{S}_a\| \stackrel{dist}{=} c_p \left\| \frac{1}{n^{1/p}} \sum_{i=1}^n \theta_i x_i \right\| .$$

To compare $E\|S_a\|$ with $E\|\tilde{S}_a\|$, one first shows $E\Sigma_{j=1}^\infty \left|\Gamma_j^{-\frac{1}{p}} - j^{-\frac{1}{p}}\right| = \Phi < \infty$ (we omit this computation). Then

$$\begin{aligned} \left| E\|S_a\| - E\|\tilde{S}_a\| \right| &\le \sum_{i=1}^k |a_i| E \sum_{j=1}^\infty \left|\Gamma_j^{-\frac{1}{p}} - j^{-\frac{1}{p}}\right| \|Y_{i,j}\| \\ &\le \sum_{i=1}^k |a_i| \cdot \Phi \cdot \max_{1\le i\le n} \|x_i\| \le k^{\frac{1}{q}} \cdot \Phi \, . \end{aligned}$$

Now,

$$E\|\tilde{S}_a\| = c_p E\| \frac{1}{n^{1/p}} \sum_{i=1}^n \theta_i x_i\| \ge \frac{c_p}{2} ST_p(X).$$

So that, if $k \le \delta(ST_p(X))^q$ for an appropriate δ, we get that $E\|S_a\|$ is approximately equal to $c_p E\|\frac{1}{n^{1/p}}\Sigma_{i=1}^n \theta_i x_i\|$ and Proposition 13.13 is proved.

□

To prove Proposition 13.14 we use Lemma 8.4. First notice that if $S = \Sigma_{i=1}^\ell x_i$ is a sum of independent randon variables taking values in some normed space, and if $\mathcal{F}_i$ is the σ-field generated by $x_1, \ldots, x_i$. Then

$$\|E(\|S\| \,|\mathcal{F}_i) - E(\|S\| \,|\mathcal{F}_{i-1})\|_\infty \le 2\|x_i\|_\infty, \; i = 2, \ldots, \ell.$$

Indeed,

$$\begin{aligned} &E(\|S\| \,|\mathcal{F}_i) - E(\|S\| \,|\mathcal{F}_{i-1}) \\ &\le E(\|\sum_{j\ne i} x_j\| \,|\mathcal{F}_i) + E(\|x_i\| \,|\mathcal{F}_i) - E(\|\sum_{j\ne i} x_i\| \,|\mathcal{F}_{i-1}) + E(\|x_i\| \,|\mathcal{F}_{i-1}) \\ &= \|x_i\| + E\|x_i\| \, , \end{aligned}$$

and a similar inequality holds for

$$E(\|S\| \,|\mathcal{F}_{i-1}) - E(\|S\| |\, \mathcal{F}_i).$$

Applying Lemma 8.4 to $\|S_a\|$ and using the estimate above we get

$$P(|\|S_a\| - E\|S_a\|| > c) \leq 2\, exp\left(\frac{-\delta_p \cdot c^q}{\|\{2 \cdot a_i \cdot j^{-1/p}\}_{i=1,j=1}^{k\ \ \infty}\|_{p,\infty}^q}\right) =$$

$$= 2exp\left(-\frac{\delta_p \cdot c^q}{2^q}\right)$$

and Proposition 13.14 follows now from Proposition 13.13.b.

□

13.16. One of the main problems in the local theory of normed spaces is to find the structure of the nicely complemented finite dimensional subspaces of a general normed space. In this connection one may wonder whether one can choose the copies of the $\ell_{p_X}^n$'s and $\ell_{q_X}^n$'s in X to be uniformly complemented. The examples of L_∞ and L_1 show that this is not the case. Moreover, a recent example of Pisier [Pis3] shows the existence of a Banach space X which admits uniformly complemented $\ell_p^n - s$ for no $1 \leq p \leq \infty$. We conclude this chapter with a proposition showing that under some additional assumptions one gets a positive result. We refer the reader to Chapter 9 for the definition of the projections Rad_n and recall that if $sup_n\|Rad_n\| < \infty$ on X then for each $2 \leq q \leq \infty$

$$C_q(X) \leq T_p(X^*) \leq sup_n\|Rad_n\| C_q(X) \qquad (\frac{1}{p} + \frac{1}{q} = 1)\ .$$

In particular $\frac{1}{p_X} + \frac{1}{q_X} = 1$.

PROPOSITION: *Let X be an infinite dimensional Banach space. Assume $sup_n\|Rad_n\| < \infty$ on X and assume in addition that X^* is of type p_{X^*} (i.e., the sup in the definition of p_{X^*} is attained). Then, for some $K < \infty$ and for each $\varepsilon > 0$ and positive integer k, X contains a K- complemented subspace which is $(1+\varepsilon)$-isomorphic to $\ell_{q_X}^k$.*

PROOF: Let $p = p_X$, $q = q_X$. By Theorem 13.2, X contains for each n and $\varepsilon > 0$ a sequence $e_1, \ldots, e_n$ with

$$\left(\sum_{i=1}^n |a_i|^q\right)^{1/q} \leq \|\sum_{i=1}^n a_i e_i\| \leq \left(1+\varepsilon\right)\left(\sum |a_i|^q\right)^{1/q}.$$

It follows from the Hahn-Banach theorem that one can find $x_1^*, \ldots, x_n^*$ in X^* with $\|x_i^*\| \leq 1$ and $x_i^*(e_j) = \delta_{ij}$, $1 \leq i,\ j \leq n$. In particular for $i \neq j$, $1 \leq \|x_i^* - x_j^*\| \leq 2$. Proposition 11.10 implies now that for each k, if n is large enough, there exists a subsequence $x_{i_1}^*, \ldots, x_{i_{2k}}^*$ such that the sequence

$$y_j^* = x_{i_{2j}}^* - x_{i_{2j-1}}^* \quad j = 1, \ldots, k$$

is 3-unconditional. Let $d_j = e_{i_{2j}}$ $j = 1, \dots, k$, then $y_i^*(d_j) = \delta_{ij}$ $1 \le i, j \le n$. Define a projection P from X onto $span\{d_j\}_{j=1}^k$ by

$$Px = \sum_{j=1}^{k} y_j^*(x) d_j .$$

Then,

$$\begin{aligned}
\|Px\| &\le (1+\varepsilon)\left(\sum_{j=1}^{k} |y_j^*(x)|^q\right)^{\frac{1}{q}} \\
&= (1+\varepsilon) sup\{|\sum_{j=1}^{k} a_j y_j^*(x)|;\ \sum |a_j|^p = 1\} \\
&\le 3(1+\varepsilon)\|x\| sup\{\int_0^1 \left\|\sum_{j=1}^{k} r_j(t) a_j y_j^*\right\| dt;\ \sum |a_j|^p = 1\} \\
&\le 3(1+\varepsilon)\|x\| T_p(X^*)\ sup\{(\sum_{j=1}^{k} |a_j|^p \|y_j^*\|^p)^{1/p};\ \sum |a_j|^p = 1\} \\
&\le 6(1+\varepsilon) T_p(X^*) \|x\|
\end{aligned}$$

and

$$\|P\| \le 6(1+\varepsilon) T_p(X^*) .$$

□

14. THE RADEMACHER PROJECTION

In this chapter we return to the subject of estimating the norm of Rad_n – the natural projection onto the subspace $\{\Sigma_{i=1}^n r_i(t)x_i\ ;\ \ x_i \in X\}$ of $L_2(X)$. The main theorem here (14.5), due to Pisier, states that $sup_{n<\infty}\|Rad_n\| < \infty$ if and only if ℓ_1 is not finitely representable in X. Towards the end of the chapter we give also an estimate for $\|Rad_n\|$ good for any finite dimensional normed space. In this chapter it will be more convenient for us to work with complex scalars so we assume all spaces are over the complex field $\mathbb{C}$. (The results are easily seen to hold also in the real case).

14.1. Given two functions $f \in L_2(\{-1,1\}^n, X)$ and $g \in L_2(\{-1,1\}^n,\ \mathbb{C})$ we may form their convolution

$$f * g(\varepsilon) = 2^{-n} \sum_{\delta \in \{-1,1\}^n} f(\varepsilon \cdot \delta) g(\delta) \qquad \varepsilon \in \{-1,1\}^n$$

where $\varepsilon \cdot \delta$ is the coordinate by coordinate multiplication. Note that if $f = \sum_{A \subseteq \{1,\ldots,n\}} w_A x_A$, $g = \sum_{\subseteq\{1,\ldots,n\}} w_A \alpha_A$, where $\{w_A\}_{A\subseteq\{1,\ldots,n\}}$ are the Walsh functions, then

$$f * g = \sum_{A \subseteq \{1,\ldots,n\}} w_A \alpha_A x_A \ .$$

Note that the norm of the operator $f \to f * g$ is smaller than or equal to $\|g\|_{L_1}$.

For $t \geq 0$ let

$$g_t(\varepsilon) = \prod_{i=1}^n (1 + e^{-t} \cdot \varepsilon_i) \ , \quad \varepsilon = (\varepsilon_1, \ldots, \varepsilon_n) \in \{-1,1\}^n$$

and let

$$S_t f = f * g_t \quad \text{for} \quad f \in L_2(\{-1,1\}^n, X) \ .$$

Then $\{S_t\}_{t\geq 0}$ is a semi-group of norm one operators on $L_2(X)$ (check). For $j = 1, \ldots, n$, let E_j be the conditional expectation operator onto the σ-field generated by the functions independent of the j-th component, i.e.,

$$E_j f(\varepsilon_1, \ldots, \varepsilon_n) = \frac{1}{2}(f(\varepsilon_1, \ldots, 1, \ldots, \varepsilon_n) + f(\varepsilon_1, \ldots, -1, \ldots, \varepsilon_n))$$

where the 1 and -1 stand in the j-th place. It is easily checked that

$$S_t = \prod_{i=1}^n (E_i + e^{-t}(I - E_i)) \ .$$

One can write

$$S_t = \sum_{k=0}^n e^{-kt} W_k \ .$$

Note that $W_1 = \Sigma_{i=1}^n (I - E_i)\Pi_{j\neq i} E_j$ is no other than Rad_n.

The main step in the proof that if ℓ_1 is not finitely representable in X then $sup_n \|Rad_n\| < \infty$, consists of showing that, in that case, S_t extends to an analytic semigroup $\{S_\xi\}_{\xi\in V}$ in some sector

$$V = \{z \in \mathbb{C} \ ; \ Re\ z \geq 0, \ |arg\ z| < \Phi\} ,$$

with $sup_{\xi\in V}\|S_\xi\| \leq K$, for some $0 < \Phi$, $K < \infty$ independent of n. Then it is quite easy to complete the proof by showing that W_1 (the coefficient of $e^{-\xi}$ in the expansion of S_ξ to power series in $e^{-\xi}$) is bounded by a bound independent of n.

We begin with two lemmas.

14.2. LEMMA: *If ℓ_1 is not finitely representable in X then there exists an integer N and a real number $0 < \rho < 2$ such that for all N-tuples $P_1, \ldots, P_N$ of commuting norm one projections on X one has*

$$\left\| \prod_{k=1}^N (I - P_k) \right\| \leq \rho^N .$$

PROOF: If this is not the case then for all N and $\varepsilon > 0$ one can find N norm one commuting projections $P_1, \ldots, P_N$ and a norm one $x \in X$ such that

$$\left\| \prod_{k=1}^N (I - P_k)x \right\| > 2^N - \varepsilon .$$

We are going to show that in that case $x, P_1x, \ldots, P_Nx$ is nicely equivalent to the unit vector basis of ℓ_1^{N+1}.

First notice that for all $\varepsilon_1, \ldots, \varepsilon_N = \pm 1$

$$\left\| \sum_{k=1}^N (I + \varepsilon_k P_k)x \right\| \geq 2^N - 2^N\varepsilon . \qquad (*)$$

Indeed, let

$$A = \{k; \varepsilon_k = -1\} , \quad B = \{k; \varepsilon_k = 1\}$$

and for $C \subseteq \{1, \ldots, N\}$ put

$$P_C = \prod_{k\in C} P_k \quad (P_\emptyset = I) .$$

Then,

$$\begin{aligned} \prod_{k=1}^N (I - P_k) &= \prod_{k\in B} (I - P_k) \prod_{k\in A} (I - P_k) \\ &= \sum_{C\subseteq B} (-1)^{|C|} P_C \prod_{k\in A} (I - P_k) \end{aligned}$$

and

$$(-1)^{|B|}P_B\prod_{k\in A}(I-P_k)=\prod_{k=1}^{N}(I-P_k)-\sum_{\substack{C\subseteq B\\ C\neq B}}(-1)^C P_C\prod_{k\in A}(I-P_k)\ .$$

Consequently,

$$\|P_B\prod_{k\in A}(I-P_k)x\|\geq 2^N-\varepsilon-(2^{|B|}-1)2^{|A|}=2^{|A|}-\varepsilon$$

and

$$\begin{aligned}\|\prod_{k\in B}(I+P_k)\prod_{k\in A}(I-P_k)x\| &\geq \|P_B\prod_{k\in B}(I+P_k)\prod_{k\in A}(I-P_k)x\|\\ &\geq 2^{|B|}\|P_B\prod_{k\in A}(I-P_k)x\|\geq 2^{|B|}(2^{|A|}-\varepsilon)\\ &\geq 2^N(1-\varepsilon)\end{aligned}$$

which proves $(*)$.

The vector $\Pi_{k=1}^N(I+\varepsilon_kP_k)x$ is equal to $x+\Sigma_{k=1}^N\varepsilon_kP_kx$ plus 2^N-N-1 vectors each of norm at most one. It follows that, for all $\varepsilon_1,\ldots,\varepsilon_N=\pm1$,

$$\|x+\sum_{k=1}^N\varepsilon_kP_kx\|\geq 2^N(1-\varepsilon)-2^N+N+1=N+1-\varepsilon\cdot 2^N\ .$$

To evaluate $\|\Sigma_{k=0}^N\alpha_kP_kx\|$ from below for general real coefficients, put $\varepsilon_k=sign\ \alpha_k$, $k=0,\ldots,N$, then

$$\begin{aligned}\left\|\sum_{k=0}^N\alpha_kP_kx\right\| &= \left\|\sum_{k=0}^N\varepsilon_k|\alpha_k|P_kx\right\|\\ &\geq\sum_{k=0}^N|\alpha_k|\left\|\sum_{k=0}^N\varepsilon_kP_kx\right\|-\left\|\sum_{k=0}^N\varepsilon_k(\sum_{\ell=0}^N|\alpha_\ell|-|\alpha_k|)P_kx\right\|\\ &\geq(N+1-\varepsilon\cdot2^N)\sum_{k=0}^N|\alpha_k|-\sum_{k=0}^N\left(\sum_{\ell=0}^N|\alpha_\ell|-|\alpha_k|\right)=(1-\varepsilon\cdot2^N)\sum_{k=0}^N|\alpha_k|\end{aligned}$$

that is, if $\varepsilon\cdot2^N<1$, $x,P_1x,\ldots,P_Nx$ is $\frac{1}{1-\varepsilon\cdot2^N}$-equivalent to the unit vector basis of ℓ_1^{N+1} over the reals. This means that ℓ_1, over the reals, is finitely representable in X. Since this implies that complex ℓ_1 is finitely representable in X(check), the proof is complete.

□

14.3. Lemma 14.2 implies in particular that if ℓ_1 is not finitely representable in X then there exists a $\rho<2$ and an $M<\infty$ such that for all n and all n-tuples of commuting norm one projections $P_1,\ldots,P_n$,

$$\left\|\prod_{k=1}^n(I-P_k)\right\|\leq M\cdot\rho^n\ .$$

Given such an n-tuple of commuting norm one projections $P_1, \ldots, P_n$, we define a semigroup $\{S_t\}_{t \geq 0}$ of operators by

$$S_t = \prod_{j=1}^{n} (P_j + e^{-t}(I - P_j)) \,, \quad t \geq 0 \,.$$

LEMMA: *Assume ℓ_1 is not finitely representable in X. Then there are $\rho < 2$ and $M < \infty$ such that, given any family $P_1, \ldots, P_n$ of commuting norm one projections and forming the semigroup $\{S_t\}_{t \geq 0}$ as above, there exists a norm $|\cdot|$ on X satisfying*

(a) $|S_t x| \leq |x|$ *for all* $t \geq 0$ *and* $x \in X$

(b) $|(I - S_t)x| \leq \rho |x|$ *for all* $t \geq 0$ *and* $x \in X$ *and*

(c) $\|x\| \leq |x| \leq M \|x\|$ *for all* $x \in X$.

PROOF: The operator S_t is a convex combination of projections from the set

$$C = \left\{ \prod_{k \in A} P_k, \quad A \subseteq \{1, \ldots, n\} \right\} \,.$$

Indeed,

$$\begin{aligned} S_t &= \prod_{j=1}^{n} (e^{-t} I + (1 - e^{-t}) P_j) \\ &= \sum_{k=0}^{n} e^{-(n-k)t} (1 - e^{-t}) \sum_{|A|=k} \prod_{j \in A} P_j \end{aligned}$$

and

$$\sum_{k=0}^{n} e^{-(n-k)t} (1 - e^{-t}) \binom{n}{k} = 1 \,.$$

It follows that $\|S_t\| \leq 1$.

Convexity considerations imply also that

$$\left\| \prod_{j=1}^{\ell} (I - S_{t_j}) \right\| \leq M \cdot \rho^{\ell}$$

for all $t_1, \ldots, t_\ell \geq 0$. One way to see this is to introduce, for $1 \leq j \leq \ell$, $1 \leq k \leq n$, independent random variables, $\xi_{j,k}(\omega)$, with

$$\begin{aligned} P(\xi_{j,k}(\omega) = 1) &= e^{-t_j} \\ P(\xi_{j,k}(\omega) = 0) &= 1 - e^{-t_j} \,. \end{aligned}$$

Then,

$$\Pi_j(\omega) = \prod_{k=1}^{n} (P_k + \xi_{j,k}(\omega)(I - P_k)) = \prod_{\{k, \xi_{j,k} = 0\}} P_k$$

and

$$S_{t_j} = \int \Pi_j(\omega) dP(\omega) \ , \quad j = 1, \ldots, \ell \ .$$

We get that, for each ω,

$$\left\| \prod_{j=1}^{\ell} (1 - \Pi_j(\omega)) \right\| \leq M \cdot \rho^{\ell}$$

(note, for each ω, $\Pi_j(\omega)$, $j = 1, \ldots, n$ is a family of commuting norm one projections so that the remark before the statement of the lemma applies to it). Consequently,

$$\begin{aligned} \left\| \prod_{j=1}^{\ell} (I - S_{t_j}) \right\| &= \left\| \prod_{j=1}^{\ell} \int (I - \Pi_j(\omega)) dP(\omega) \right\| \\ &= \left\| \int \prod_{j=1}^{\ell} (I - \Pi_j(\omega)) dP(\omega) \right\| \\ &\leq \int \left\| \prod_{j=1}^{\ell} (I - \Pi_j(\omega)) \right\| dP(\omega) \leq M \cdot \rho^{\ell} \ . \end{aligned}$$

We introduce a new norm on X by

$$|x| = sup \left\{ \rho^{-\ell} \left\| \prod_{j=1}^{\ell} (I - S_{t_j}) \prod_{k=1}^{m} S_{r_k} x \right\| \right\}$$

where the supremum is taken over all ℓ, m, $t_1, \ldots, t_\ell$ and $r_1, \ldots, r_m$. The three required properties follow immediately from the two inequalities

$$\|S_t\| \leq 1 \quad \text{and} \quad \| \prod_{j=1} (I - S_{t_j}) \| \leq M \cdot \rho^{\ell} \ .$$

□

14.4. The main tool in the proof of Theorem 14.5 below is an analytic extension theorem, due to Beurling and Kato, for strongly continuous semigroups of operators. We recall that a semigroup $\{S_t\}_{t \geq 0}$ of operators on X is said to be *strongly continuous* if $S_t x \underset{t \to 0}{\longrightarrow} x$ for every $x \in X$. If V is some open cone in $\mathbb{C}$ a semigroup $\{S_\xi\}_{\xi \in V}$ is said to be *analytic* if $x^*(S_\xi x)$ is an analytic function in V for all $x \in X$, $x^* \in X^*$.

THEOREM: *For all $1 \leq \rho < 2$ there exist constants $\phi > 0$ and $K < \infty$ such that if $\{S_t\}_{t \geq 0}$ is a strongly continuous semigroup of contractions on some complex Banach space satisfying $sup_{t \geq 0} \|I - S_t\| \leq \rho < 2$. Then $\{S_t\}_{t \geq 0}$ admits an analytic extension $\{S_\xi\}_{\xi \in V}$ where*

$$V = \{z \in \mathbb{C}; \ \ Re \ z \geq 0, \ \ |arg \ z| < \phi\}$$

and $\{S_\xi\}_{\xi \in V}$ satisfies

$$\sup_{\xi \in V} \|S_\xi\| \leq K \ .$$

We postpone the proof to Appendix IV.

14.5. THEOREM. *Let X be a (complex) Banach space. Then ℓ_1 is not finitely representable in X if and only if $sup_n \|Rad_n\|_{L_2(X)} < \infty$.*

PROOF: By Lemma 9.10 (see also 9.11), $sup_{n,m} \|Rad_n\|_{L_2(\ell_1^m)} = \infty$. (It is actually possible to show that $\|Rad_n\|_{L_2(\ell_1^n)} \approx \sqrt{\log n}$.) It follows that if $sup_n \|Rad_n\|_{L_2(X)} < \infty$ then ℓ_1 is not finitely representable in X.

If ℓ_1 is not finitely representable in $L_2(X)$then, defining, as in the beginning of this chapter

$$S_t = \prod_{i=1}^{n} (E_i + e^{-t}(I - E_i)) ,$$

we get by Lemmas 14.2, 14.3 and Theorem 14.4 that $\{S_t\}_{t \geq 0}$ admits an analytic extension $\{S_\xi\}_{\xi \in V}$ to some sector

$$V = \{z \in \mathbb{C}; \quad Re\, z \geq 0, \quad |arg\, z| < \phi\}$$

with

$$\sup_{\xi \in V} \|S_\xi\| \leq K$$

where $\phi > 0$ and K are independent of n.

Necessarily,

$$S_\xi = \prod_{j=1}^{n} (P_j + e^{-\xi}(I - E_j)) = \sum_{k=0}^{n} e^{-k\xi} W_k$$

(since the right hand side defines an analytic semigroup which coincides with S_t for $t \in \mathbb{R}$, $t \geq 0$). Recall that $W_1 = Rad_n$.

For $a = \pi / tg\phi$, $a + ib \in V$ for all $-\pi \leq b \leq \pi$. Also, it is easily checked that

$$W_1 = \frac{e^a}{2\pi} \int_{-\pi}^{\pi} S_{a+ib} e^{ib} db .$$

It follows that

$$\|Rad_n\| = \|W_1\| \leq e^a \cdot K = e^{\pi/tg\phi} \cdot K .$$

It remains to check that if ℓ_1 is not finitely representable in X then it is not finitely representable in $L_2(\{-1,1\}^n, X)$ uniformly in n, i.e., that it is not finitely representable in $L_2([0,1], X)$. This follows from the fact (Theorem 13.2) that if ℓ_1 is not finitely representable in X then X has some type $p > 1$. Then, by 9.12, $L_2([0,1], X)$ has the same type and consequently ℓ_1 is not finitely representable in $L_2([0,1], X)$.

□

14.6. We now state a theorem estimating $\|Rad_n\|$ for a general finite dimensional space.

THEOREM: *Let X be a finite dimensional normed space of dimension k. Then, for all n, m*

$$\|Rad_n\|_{L_2(\{-1,1\}^m,X)} \le (e+1)log(d(X,\ell_2^k)+1) .$$

In particular,

$$\|Rad_n\|_{L_2(\{-1,1\}^m,X)} \le K \cdot log\ k$$

for some absolute constant K.

PROOF: Note that $Rad_n f = f * g$ where $g = \Sigma_{i=1}^n r_i$. For any g, $\|f * g\|_{L_2(X)} \le \|f\|_{L_2(X)} \cdot \|g\|_{L_1}$. If X is a Hilbert space and $g = \Sigma c_A w_A$ we also have $\|f * g\|_{L_2(X)} \le \|f\|_{L_2(X)} \cdot max|c_A|$. It follows that, *in any space* X

$$\|f * g\|_{L_2(X)} \le \|f\|_{L_2(X)} \cdot max|c_A| \cdot d(X, \ell_2^{dim\ X}) .$$

For $-1 \le \alpha \le 1$ let $g_\alpha(\varepsilon) = \Pi_{i=1}^n(1+\alpha\varepsilon_i)$ then $g_\alpha \ge 0$, $\int g_\alpha = 1$, $g_\alpha = \Sigma_{k=0}^n \alpha^k \Sigma_{|A|=k} w_A$. For $\mu \in (C[-1,1])^*$ let $g = \int_{-1}^1 g_\alpha d\mu(\alpha)$. Then

$$\|f * g\|_{L_2(X)} \le \int_{-1}^1 \|f * g_\alpha\|_{L_2(X)} d\mu(\alpha) \le \|f\|_{L_2(X)}\|\mu\| .$$

That is, the operator $T_g f = f * g$ has norm $\le \|\mu\|$. Since $g = \Sigma_{k=0}^n (\int_{-1}^1 \alpha^k d\mu(\alpha))\Sigma_{|A|=k} w_A$ we get that if μ satisfies

$$(a)\quad \int_{-1}^1 \alpha d\mu(\alpha) = 1 \quad \text{and} \quad (b)\quad \max_{\substack{0\le k\le n \\ k\ne 1}} \left|\int_{-1}^1 \alpha^k d\mu(\alpha)\right| = \delta$$

then

$$\|Rad_n\| = \|T_{(\Sigma_{i=1}^n r_i)}\| \le \|T_g\| + \|T_{(g-\Sigma_{i=1}^n r_i)}\| \le \|\mu\| + \delta \cdot d(X, \ell_2^k) .$$

Bernstein's inequality (see e.g. [Zy]) states that for a degree ℓ polynomial p on $[-1,1]$

$$|p'(0)| \le \ell \max_{-1\le t\le 1} |p(t)| .$$

It follows from the Hahn-Banach theorem that there exists a $\mu \in (C[-1,1])^*$ with $\|\mu\| \le \ell$ and with $\int_{-1}^1 t d\mu(t) = 1$ and $\int_{-1}^1 t^k d\mu(t) = 0$ for $k = 0,2,3,\ldots,\ell$. Let $a > 1$ and define

$$\nu(A) = a\mu(aA).$$

Then

$$\|\nu\| \le a \cdot \ell, \quad \int_{-1}^1 t d\nu(t) = 1 \quad \text{and} \quad \int_{-1}^1 t^k d\nu(t) = 0 \quad \text{for} \quad k = 0,2,\ldots,\ell .$$

For $k > \ell$ we have

$$\int_{-1}^1 t^k d\nu(t) = a\int_{-1}^1 (a^{-1}\cdot t)^k d\mu(t) = a^{1-k}\int_{-1}^1 t^k d\mu(t) ,$$

so that

$$\left|\int_{-1}^{1} t^k d\nu(t)\right| \leq a^{1-k} \cdot \ell \leq a^{-\ell} \cdot \ell$$

and we get

$$\|Rad_n\| = \|T_{(\sum_{i=1}^n r_i)}\| \leq a \cdot \ell + a^{-\ell} \cdot \ell \cdot d(X, \ell_2^k) \ .$$

Taking $a = e$ and $\ell = [log\ d(X, \ell_2^k)] + 1$ we get the desired result.

□

14.7. We conclude the chapter with a corollary to Theorem 14.5 which shows that the ℓ_1, ℓ_∞ example in 9.11 is essentially the only example to the case where duality between type of a space and cotype of its dual does not hold.

COROLLARY: *Let X be a Banach space with $p_X > 1$. Then $p_X^{-1} + q_{X^*}^{-1} = 1$.*

PROOF: If $p_X > 1$ then ℓ_1 is not finitely representable in X so that by Theorem 14.5 $sup_n \|Rad_n\|_{L_2(X)} < \infty$ and the corollary follows from Lemma 9.10.

□

15. PROJECTIONS ON RANDOM EUCLIDEAN SUBSPACES OF FINITE DIMENSIONAL NORMED SPACES

In this chapter we combine results from Chapters 4 and 14 to prove the existence of nicely complemented almost euclidean subspaces in a large class of normed spaces (Theorem 15.10 below). After some introductory results, we bring in 15.4 the general scheme of the proof.

15.1. We start by introducing two norms on the space $B(\ell_2^n, X)$ of operators from ℓ_2^n into a normed space X. In both cases we shall omit the subscript n from the notation.

15.1.1. *The ℓ-norm:* For $A : \ell_2^n \to X$ define

$$\ell(A) = \sqrt{n}\left(\int_{S^{n-1}} \|Ax\|^2 d\mu\right)^{1/2}$$

where μ is the normalized Haar measure on S^{n-1} – the unit sphere of ℓ_2^n. Recall (cf. 9.5.1) that one can also write

$$\ell(A) = (E\|\sum_{i=1}^{n} g_i A e_i\|^2)^{1/2}$$

where $(e_i)_{i=1}^n$ is the unit vector basis of ℓ_2^n and $(g_i)_{i=1}^n$ are independent, symmetric, gaussian random variables normalized in L_2.

15.1.2. *The r-norms:* Fix an orthonormal basis $u = (u_1, \dots, u_n)$ in ℓ_2^n and define

$$r_u(A) = \left(\int_0^1 \|\sum_{i=1}^{n} r_i(t) A u_i\|^2 dt\right)^{1/2}$$

where $(r_i)_{i=1}^n$ are the Rademacher functions.

EXERCISE: There is an obvious one to one correspondence between the set of all orthonormal bases and $O(n)$ – the orthogonal group. Show

$$\left(\int_{O(n)} r_u(A)^2 d\mu(u)\right)^{1/2} = \ell(A)$$

where μ is the normalized Haar measure on $O(n)$.

The norms ℓ and r_u are not ideal norms in the usual sense of this notion (see [Pie]), however they share some of the properties of ideal norms. In particular

(i) $\ell(A) = \|A\|$ for any rank 1 operator (use the second definition of $\ell(A)$). And,

(ii) $\ell(A\ B\ C) \le \|A\| \cdot \ell(B) \cdot \|C\|$ for any normed spaces X, Y and operators $A : X \to Y$, $B : \ell_2^n \to X$, $C : \ell_2^n \to \ell_2^n$ (first check it for unitary C then use the fact that any norm one operator on ℓ_2^n is a convex combination of unitary operators).

We remark in passing that an ideal norm is a norm, defined on the space $B(V,X)$ of operators from V to X for *any* two finite dimensional normed spaces V and X, which satisfy (i) and (ii) for *any* diagram

$$U \xrightarrow{C} V \xrightarrow{B} X \xrightarrow{A} Y .$$

15.1.3. *The dual norm:* Given any two finite dimensional normed spaces X and Y and a norm i on $B(X,Y)$, one defines a norm i^* on $B(Y,X)$ by

$$i^*(B) = sup\{trace\ BA; \quad A \in B(X,Y), \quad i(A) \leq 1\} .$$

Clearly, $(B(Y,X), i^*)$ is the dual space to $(B(X,Y), i)$ with duality $< A, B >= trace\ BA$.

15.2. THEOREM: *Let X and Y be n-dimensional normed spaces. Let α be a norm on $B(X,Y)$ and let α^* be the dual norm. Then there exists an operator $W : X \to Y$ such that $\alpha(W) \cdot \alpha^*(W^{-1}) = n$.*

REMARKS:

1) Note that for any W, $n = trace\ W \cdot W^{-1} \leq \alpha(W) \cdot \alpha^*(W^{-1})$.

2) One can prove F. John's Theorem 3.3 as a corollary to Theorem 15.2, see [Lew] and [Pe].

PROOF OF THEOREM 15.2: Let W be the operator in $B(X,Y)$ for which the maximum

$$max\{|det\ U|; \quad U : X \to Y, \quad \alpha(U) = 1\}$$

is attained (the set $U : X \to Y$, $\alpha(U) = 1$ is compact). We shall show that $\alpha^*(W^{-1}) = n$ (W is clearly invertible).

Let $U : X \to Y$ be arbitrary and consider

$$\beta = |det(\alpha(W + tU)^{-1} \cdot (W + tU))|$$

which is well defined for small $|t|$. By the choice of W, $\beta \leq |det\ W|$. On the other hand

$$\beta = \alpha(W + tU)^{-n} \cdot |det\ W| \cdot |det(I + tW^{-1}U)| .$$

Hence

$$|det(I + tW^{-1}U)|^{1/n} \leq \alpha(W + tU) \leq \alpha(W) + t\alpha(U) = 1 + t\alpha(U) .$$

Using the Jordan canonical form it is easily seen that

$$\frac{1}{n} tr(W^{-1}U) = \lim_{t \to 0} \frac{det(I + tW^{-1}U)^{1/n} - 1}{t} \leq \alpha(U) .$$

Thus, $\alpha^*(W^{-1}) \leq n$.

□

15.3. The next lemma relates the r_u^* norm of T to the r_u norm of its adjoint.

LEMMA: *Let $T: X \to \ell_2^n$ and let $u = (u_1, \ldots, u_n)$ be any orthonormal basis in ℓ_2^n. Then*

$$r_u^*(T) \le r_u(T^*) \le \|Rad_n\|_{L_2(X)} r_u^*(T) .$$

PROOF: By the orthogonality of the Rademacher functions, for any $S: \ell_2^n \to X$

$$\begin{aligned} trace\ TS &= \sum_{i=1}^n < TSu_i, u_i > = \sum_{i=1}^n < Su_i, T^* u_i > \\ &= \int_0^1 < \sum_{i=1}^n r_i(t) Su_i, \sum_{i=1}^n r_i(t) T^* u_i > \\ &\le r_u(S) r_u(T^*) , \end{aligned}$$

and we conclude that

$$r_u^*(T) \le r_u(T^*) .$$

To prove the other inequality recall the duality between $L_2(X)$ and $L_2(X^*)$. It follows that for some $\{y_i\}_{i=1}^n$, $\{y_A\}_{\substack{|A| \ne 1 \\ A \subseteq \{1,\ldots,n\}}}$ with

$$\left\| \sum_{i=1}^n y_i r_i(t) + \sum_{|A| \ne 1} y_A \omega_A(t) \right\|_{L_2(X)} \le 1 ,$$

$$r_u(T^*) = \left\| \sum_{i=1}^n r_i(t) T^* u_i \right\|_{L_2(X)} = \sum_{i=1}^n < T^* u_i, y_i > .$$

Define $S: \ell_2^n \to X$ by $Su_i = y_i$. Then

$$r_u(S) r_u^*(T) \ge trace\ TS = \sum_{i=1}^n < u_i, TSu_i > = \sum_{i=1}^n < T^* u_i, y_i > = r_u(T^*)$$

and

$$r_u(S) = \left\| \sum_{i=1}^n r_i(t) y_i \right\|_{L_2(X)} \le \|Rad_n\|_{L_2(X)} .$$

□

15.4. As a corollary to Theorem 15.2 and Lemma 15.3 we get that there exists a $T: \ell_2^n \to X$ with $r_u(T) r_u(T^{-1*}) \le \|Rad_n\| \cdot n$.

The result of Appendix II implies that for any $q > 2$ and $x_1, \ldots, x_n \in X$,

$$\left(E \left\| \sum_{i=1}^n g_i x_i \right\|^2 \right)^{1/2} \le K \left(\int_0^1 \left\| \sum_{i=1}^n r_i(t) x_i \right\|^2 \right)^{1/2}$$

where K depends on q and $C_q(X)$ – the cotype q constant of X – only. A similar inequality holds for X^*. We conclude that

15.4.1.

$$\ell(T)\ell(T^{-1*}) \le K\|Rad_n\| \cdot n$$

where K depends only on q, q^*, $C_q(X)$ and $C_{q^*}(X^*)$. Moreover, using Exercise 15.1.2, one can easily get 15.4.1 with a universal constant K.

We now introduce an euclidean norm on X by taking its unit ball to be the image by T of the unit ball of ℓ_2^n. Using the method of Chapter 4 and in particular Proposition 4.6, we would like to find a large dimensional almost euclidean subspace of X which is nicely complemented. The quantities M_r and $\frac{1}{\sqrt{n}}\ell(T)$ are closely related (the first is the median and the second in the quadratic mean of $\|Tx\|$ on S^{n-1}). Instead of relating everything to M_r and using Chapter 4 we prefer to work with $\ell(T)$ and imitate the method of Chapter 4.

The production of the complemented euclidean spaces will be done in three stages. Let $X \subset Y$, $dim\ X = n$ and let $T: \ \ell_2^n \to X$ be as above. In the first step (Lemma 15.5) we show that $T^{-1}: \ X \to \ell_2^n$ can be extended to $S: \ Y \to \ell_2^n$ in such a manner as to preserve the r_u^* norm. Then one gets the same estimate for $\ell(T)\ell(S^*)$ as in 15.4.1. The second step (Corollary 15.8) is to pass to a large dimensional subspace E of ℓ_2^n on which we have good estimates for $\|T_{|E}\|$ and $\|S^*_{|E}\|$ in terms of the ℓ norms of T and S^* respectively. The last step (Lemma 15.9) is an imitation of Proposition 4.6 which shows that one can produce a good projection onto TE.

15.5. LEMMA: *Let α be a norm defined on $B(X,Y)$ for some finite dimensional normed space X and all finite dimensional normed spaces Y, which is injective in the sense that if $Y \subset Z$, with i the inclusion map, then for all $T: \ X \to Y$, $\alpha(T) = \alpha(iT)$. Then for any $Y \subset Z$ and any $R: \ Y \to X$, R admits an extension $\tilde{R}: \ Z \to X$ with $\alpha^*(\tilde{R}) = \alpha^*(R)$.*

Note that r_u and ℓ are injective.

PROOF OF LEMMA 15.5: The injectivity of α implies that $(B(X,Y),\alpha)$ is naturally a subspace of $(B(X,Z),\alpha)$. Consider the functional f on $(B(X,Y),\alpha)$ defined by $f(T) =$ *trace* RT whose norm is $\alpha^*(R)$. By the Hahn-Banach theorem one can extend f to a functional $\tilde{f}$ on $(B(X,Z),\alpha)$ with the same norm. Any such functional is given by an operator $\tilde{R}: \ Z \to X$ acting by $\tilde{f}(T) =$ *trace* $(\tilde{R}T)$. It is easy to check that $\tilde{R}$ extends R. Finally, $\alpha^*(\tilde{R}) = \|\tilde{f}\| = \|f\| = \alpha^*(R)$.

□

The second step (the passage to a subspace on which we have good estimates for both $\|T_{|F}\|$ and $\|S^*_{|F}\|$) will be done in two propositions. The proof of the first one is deterministic and of the second one probabilistic. We recall that $\beta_2(X,k)$ is the gaussian cotype 2 constant of X on k vectors (see 9.4). Clearly (see 9.12), $\beta_2(L_2(X),k) = \beta_2(X,k)$ so one gets easily that

$$\beta_2((B(\ell_2^n, X),\ell),k) = \beta_2(X,k) \ .$$

15.6. PROPOSITION: *Let $U : \ell_2^n \to X$ be any operator. Then for every $1 \le k \le n$ there exists an $E \subseteq \ell_2^n$ with $\dim E \ge n-k$ and*

$$\|U_{|E}\| \le \frac{1}{\sqrt{k}}\beta_2(X,k)\ell(U) .$$

PROOF: First notice that if $\{P_i\}_{i=1}^k$ are pairwise orthogonal projections in ℓ_2^n then

15.6.1.
$$\min_{1\le i\le k} \ell(UP_i) \le \frac{1}{\sqrt{k}}\beta_2(X,k)\ell(U) .$$

Indeed,

$$k \min_{1\le i\le k} \ell(UP_i)^2 \le \sum_{i=1}^k \ell(UP_i)^2 \le \beta_2((B(\ell_2^n,X),\ell),k)^2 \int_0^1 \ell(\sum_{i=1}^k r_i(t)UP_i)^2 dt$$
$$\le \beta_2(X,k)^2\ell(U)\int_0^1 \left\|\sum_{i=1}^k r_i(t)P_i\right\|^2 dt \le \beta_2(X,k)^2\ell(U) .$$

Let now $\{P_i\}_{i=1}^m$ be a maximal set of pairwise orthogonal projections with rank $P_i = 1$ such that, for all $1 \le i \le m$,

$$\ell(UP_i) > \frac{1}{\sqrt{k}}\beta_2(X,k)\ell(U) .$$

By 15.6.1, $m < k$. Let $F = (Im\Sigma_{i=1}^m P_i)^\perp$. Then, by the maximality of $\{P_i\}_{i=1}^m$, any rank one projection P with $Im\, P \subseteq F$ satisfies

$$\ell(UP) \le \frac{1}{\sqrt{k}}\beta_2(X,k)\ell(U) .$$

Finally,

$$\|U_{|F}\| = sup\{\|UP\|;\ P : \ell_2^n \to F,\ P \text{ a rank one orthogonal projection}\} .$$

For rank one operators the operator and the ℓ-norm coincide, hence,

$$\|U_{|F}\| \le \frac{1}{\sqrt{k}}\beta_2(X,k)\ell(U)$$

and F is the desired subspace.

□

We apply the proposition for both T and S^* (see 15.4. S is the extension of T^{-1} to Y in a manner which preserves the r_u^* norm) with $k = \frac{1}{4}n$ to obtain two subspaces F_1, F_2 such that $dim\ F_1$, $dim\ F_2 \ge \frac{3}{4}n$ and

$$\|T_{|F_1}\| \le \frac{2}{\sqrt{n}}\beta_2(X,n)\ell(T) , \quad \|S^*_{|F_2}\| \le \frac{2}{\sqrt{n}}\beta_2(Y^*,n)\ell(S^*) .$$

The subspace $F = F_1 \cap F_2$ has $dim\ F \geq \frac{1}{2}n$ and

$$\|T_{|F}\| \leq \frac{2}{\sqrt{n}}\beta_2(X,n)\ell(T) , \quad \|S^*_{|F}\| \leq \frac{2}{\sqrt{n}}\beta_2(Y^*,n)\ell(S^*) .$$

15.7. PROPOSITION: *Let $U : \ell_2^m \to X$ be any operator. Let μ be the normalized Haar measure on S^{m-1}. Then for any $C \geq 1$*

$$\mu\{x \in S^{m-1};\ \|Ux\| > \frac{3C}{\sqrt{m}}\ell(U)\} \leq 2\ exp\left\{-\frac{C^2\ell(U)^2}{\|U\|^2}\right\} .$$

PROOF: The function $x \to \|Ux\|$ is a Lipschitz function with constant $\leq \|U\|$. Let M denote the median of this function. Then

$$\sqrt{\frac{1}{2}}M \leq \mu\{x \in S^{m-1};\ \|Ux\| \geq M\}^{\frac{1}{2}} \cdot M \leq \frac{1}{\sqrt{m}}\ell(U) .$$

So $\frac{2C}{\sqrt{m}}\ell(U) \geq M$ and by Chapter 2,

$$\begin{aligned}
&\mu\{x \in S^{m-1};\ \|Ux\| > \frac{3C}{\sqrt{m}}\ell(U)\} \\
&\leq \mu\{x \in S^{m-1};\ \|Ux\| - M > \frac{C}{\sqrt{m}}\ell(U)\} \\
&\leq 2\ exp\left\{-\frac{C^2\ell(U)^2}{\|U\|^2}\right\} .
\end{aligned}$$

□

We apply this proposition to $F = \ell_2^m$, $m \geq \frac{n}{2}$ and to the operators T_F and $S^*_{|F}$ to get

$$\begin{aligned}
&\mu\{x \in F;\ \|Tx\| \leq \frac{3}{\sqrt{m}}\ell(T_{|F}) \text{ and } \|S^*x\| \leq \frac{3}{\sqrt{m}}\ell(S^*_{|F})\} \\
&\geq 1 - 2\ exp\left(\frac{-n}{4\beta_2(X,n)^2}\right) - 2\ exp\left(\frac{-n}{4\beta_2(Y^*,n)^2}\right) .
\end{aligned}$$

The method of Chapter 2 now implies

15.8. COROLLARY: *with T and S as in 15.4, there exists a subspace $E \subseteq \ell_2^n$ with*

$$dim\ E \geq c\ min\left\{\frac{n}{\beta_2(X,n)^2}, \frac{n}{\beta_2(Y^*,n)^2}\right\} ,$$

for some absolute constant $c > 0$, on which

$$\|T_{|E}\| \leq \frac{4}{\sqrt{n}}\ell(T) , \quad \|S^*_{|E}\| \leq \frac{4}{\sqrt{n}}\ell(S^*) .$$

This concludes the second step. The third step consists of a standard lemma resembling Proposition 4.6.

15.9. LEMMA: *Let $X \subseteq Y$ be Banach spaces with dim $X = n < \infty$. Let $T : \ell_2^n \to X$ be a $1-1$ operator, and let $S : Y \to \ell_2^n$ be an extension of T^{-1}. Let $F \subseteq \ell_2^n$ be any subspace. Then there exists a projection P from Y onto TF with $\|P\| \leq \|T_{|F}\| \, \|S^*_{|F}\|$.*

PROOF: Let Q be the orthogonal projection from ℓ_2^n onto F and let $P = TQS$. P is clearly a projection onto TF. Now, for any $x \in Y$,

$$\begin{aligned}\|QSx\|^2 &= (QSx, QSx) = (Sx, QSx) \\ &= < x, S^*QSx > \leq \|x\| \, \|S^*QSx\| \\ &\leq \|x\| \, \|S^*_{|F}\| \, \|QSx\| ,\end{aligned}$$

so that

$$\|QSx\| \leq \|x\| \, \|S^*_{|F}\| .$$

Finally,

$$\|Px\| = \|TQSx\| \leq \|T_{|F}\| \cdot \|QSx\| \leq \|T_{|F}\| \, \|S^*_{|F}\| \, \|x\| .$$

□

Combining the consideration in 15.4, Lemma 15.5, Corollary 15.8, Lemma 15.9 and Theorem 14.5, we get

15.10. THEOREM: *Let Y be an infinite dimensional Banach space with type $p > 1$ (i.e. $T_p(Y) < \infty$ for some $p > 1$). Then there exists a constant $C < \infty$ and, for each $\varepsilon > 0$ and positive integer k, an integer $N = N(\varepsilon, k)$ such that if $X \subseteq Y$ and dim $X \geq N$ then X contains a k-dimensional subspace Z with $d(Z, \ell_2^k) \leq 1 + \varepsilon$ and there exists a projection of norm $\leq C$ from Y onto Z.*

PROOF: The space Y must have some finite cotype otherwise it contains, by Theorem 13.2, uniform copies of ℓ_∞^n's and thus does not have any type > 1. Lemma 9.10 implies that for some $q < \infty$ and $C < \infty$, for all finite dimensional $Y' \subseteq Y$, $C_q(Y'^*) \leq C$. Also, by Theorem 14.5, $R = sup_N \|Rad_N\|_{L_2(Y)} < \infty$.

Let $X \subseteq Y$ be an N dimensional subspace and let Y' be any finite dimensional space with $X \subseteq Y' \subseteq Y$. By Theorem 15.2 we get an operator $T : \ell_2^N \to X$ with $r_u(T) r_u^*(T^{-1}) = N$ (u is any orthonormal basis). Let S be an extension of T^{-1} to an operator

$$S : Y' \longrightarrow \ell_2^N$$

with $r_u^*(S) = r_u^*(T^{-1})$ (which exists by Lemma 15.5). Then, by Lemma 15.3,

$$\begin{aligned}r_u(T) r_u(S^*) &\leq r_u(T) \|Rad_N\| r_u^*(S) \\ &\leq R r_u(T) r_u^*(T^{-1}) = R \cdot N .\end{aligned}$$

By the relation between the Rademacher and Gaussian averages (Appendix II)

15.10.1. $$\ell(T)\ell(S^*) \leq K \cdot N$$

where K depends on the cotypes and the cotype constants of Y' and Y'^*.

Corollary 15.8 now implies the existence of a subspace E of ℓ_2^N with

$$n = dim\ E \geq c \cdot min \left(\frac{N}{\beta_2(X,N)^2}, \frac{N}{\beta_2(Y^*,N)^2} \right)$$

on which

$$\|T_{|E}\| \, \|S^*_{|E}\| \leq 16 \cdot K \ .$$

This implies (since $\|S^*_{|E}\| \geq \|(T_{|TE})^{-1}\|$) that

$$d(E, \ell_2^k) \leq 16 \cdot K$$

and, by Lemma 15.9, that TE is $16 \cdot K$-complemented in Y'.

Holder's inequality implies easily that

$$\beta_2(Z,N) \leq N^{(q-2)/2q} \beta_q(Z,N)$$

for any normed space Z and any $q > 2$. Thus, since $C_q(Y'), C_q(Y'^*) \leq C$ for some q and C independent of Y', $n \to \infty$ as $N \to \infty$. An application of Theorem 4.3 produces a subspace Z of E of dimension k, where $k \to \infty$ as $n \to \infty$, which is $(1+\varepsilon)$-isomorphic to ℓ_2^k. Of course this subspace is $16 \cdot K$-complemented in E and thus $(16 \cdot K)^2$-complemented in Y'. Finally, a compactness argument, let us pass from Y' to Y.

$\square$

APPENDIX I

ISOPERIMETRIC INEQUALITIES IN RIEMANNIAN MANIFOLDS

by M. GROMOV

1. GENERALITIES ON RIEMANNIAN MANIFOLDS

1.1. Local Riemannian structures. A *Riemannian C^r-structure* g on an open subset U in the Euclidean space $\mathbb{R}^n$ is, by definition, a C^r-map $u \to g_u$, $u \in U$, which assigns to each point $u \in U$ a positive definite quadratic form (i.e. an Euclidean metric) g_u on $\mathbb{R}^n$. If $X_1, \ldots, X_n$ is a fixed basis in $\mathbb{R}^n$, then g is determined by the $\frac{n(n+1)}{2}$ scalar products $g_u(X_i, X_j)$ which are C^r-functions on U.

One defines next the g – *length* of each C^1-map $f : [0,1] \to U$ by

$$\int_0^1 (g_{f(t)}(f'(t), f'(t))^{\frac{1}{2}} dt \ ,$$

where f' stands for the derivative $\frac{df}{dt} : [0,1] \to \mathbb{R}^n$. For example, if g_u is constant in $u \in U = \mathbb{R}^n$, then the family g_u reduces to a single Euclidean structure on $\mathbb{R}^n$ and the above notion of the length agrees with the Euclidean one.

Finally, for *connected* domains U, one defines the distance $dist_g(u_0, u_1)$ between points u_0 and u_1 in U as the infimum of the lengths of C^1-*curves between u_0 and u_1*, which are C^1-maps $f : [0,1] \to U$ such that $f(0) = u_0$ and $f(1) = u_1$. This $dist_g$ is called *the Riemannian metric* associated to g and it obviously satisfies the ordinary metric axioms. Moreover, the topology associated to $dist_g$ agrees with the induced Euclidean topology in U. Observe that $dist_{g_1} = dist_{g_2}$ implies $g_1 = g_2$. Indeed, $(g_u(X,X))^{\frac{1}{2}} = \lim_{\varepsilon \to 0} \varepsilon^{-1} dist_g(u, u + \varepsilon X)$ for all $u \in U$ and $X \in \mathbb{R}^n$. However, only very special metrics on U are of the form $dist_g$ for some g. For example, no metric $dist(u_1, u_2) = \|u_1 - u_2\|$ for a *non-Euclidean* norm $\| \ \|$ on $\mathbb{R}^n$ is ever Riemannian.

EXAMPLES. (a) Let U be an open interval, say (a,b) in $\mathbb{R}^1$. Then each Riemannian structure is given by a single function g_u and the metric space (U, g_u) is isometric to the ordinary (i.e. with the metric $|u_1 - u_2|$) interval (a', b') in $\mathbb{R}^1$ for $a' = \int_a^c (g_u)^{\frac{1}{2}} du$ and $b' = \int_c^b (g_u)^{\frac{1}{2}} du$, where one may take an arbitrary fixed point in (a, b) for c.

(b) Take $\mathbb{R}^2$ with the standard Euclidean structure g_0 and let U be the unit disk in $\mathbb{R}^2$. Put $g_u = 4h(u)g_0$ for the function $h(u) = (1 - dist^2_{g_0}(u, 0))^{-2} = (1 - g_0(u, u))^{-2}$. This is called the *Poincare metric* and the metric space $(U, dist_g)$ is called the ***hyperbolic plane***. Every map $f : [0, 1) \to U$ for which $g_0(f(t), f(t)) \to 1$ for $t \to 1$ clearly has *infinite* length. This implies (by an easy argument) the *completeness* of the hyperbolic plane. The remarkable (though easy to prove) property of this g is the invariance under conformal transformations of the disk U. In fact the isometry group of $(U, dist_g)$ is (easily seen to be) locally isomorphic to the multiplicative group of unimodular 2×2 matrices.

1.2. Riemannian manifolds. A metric space $(V, dist)$ is called an *n-dimensional Riemannian C^r-manifold* if it is locally isometric to some $(U, dist_g)$. Namely, for each point $v \in V$ there are some neighbourhoods U' and $U'' \subset U'$ of v in V and a homeomorphism H of U' on an open subset $U \subset \mathbb{R}^n$ with some Riemannian C^r-metric $dist_g$ such that $dist_g(H(u_1), H(u_2)) = dist(u_1, u_2)$ for all points u_1 and u_2 in U''.

EXAMPLES. (a) Let V be a C^{r+1}-smooth n-dimensional submanifold in some Euclidean space $\mathbb{R}^q$, $q \geq n$, with the standard Euclidean metric. Recall that a subset $V \subset \mathbb{R}^q$ is a C^{r+1}-submanifold if there exists, for each $v_0 \in V$, a linear n-dimensional subspace $\mathbb{R}^n \simeq L \subset \mathbb{R}^q$, such that the orthogonal projection, say $P : V \to L$, homeomorphically maps a small neighbourhood $U' \subset V$ of v_0 onto an open subset $U \subset L$ and such that the inverse map $P^{-1} : U \to U' \subset \mathbb{R}^q$ is a C^{r+1}-map of U into $\mathbb{R}^q$. Define $dist(v_1, v_2)$, for $v_1, v_2 \in V$, by taking the infimum of the lengths (measured by the Euclidean metric in $\mathbb{R}^q \supset V$) of C^1-curves in V between v_1 and v_2. Then the resulting metric in V is Riemannian. Indeed the orthogonal projection $\mathbb{R}^q \to L \approx \mathbb{R}^n$ sends the tangent n-planes $T_v(V) \subset \mathbb{R}^q$ of V at points $v \in U'$(which are close to v_0) onto L. This brings the Euclidean structure on $T_v(V)$ down to L, say to g_u on L for $u = P(v)$, and the projection $P : U' \to U$ clearly is an isometry near v_0 for the resulting Riemannian metric on U.

This metric on $V \subset \mathbb{R}^q$ is called the *induced Riemannian* metric. The corresponding Riemannian distance in V is typically greater than the distance in the ambient space $\mathbb{R}$, since straight intervals in $\mathbb{R}^q$ with the ends in V not always lie in V.

REMARKS. (a) A deep theorem of Nash claims that every Riemannian manifold V is isometric to some manifold in $\mathbb{R}^q$ (where q must be rather large compared to $n = dim\ V$). For instance, the hyperbolic plane can be realized by a C^1-submanifold in $\mathbb{R}^3$, but no such C^2-realization exists. However, every bounded ball (or rather disk) in the hyperbolic plane is isometric to a C^∞-submanifold in $\mathbb{R}^3$. It is unknown if there is a global realization of the hyperbolic plane in $\mathbb{R}^4$, but this is possible in $\mathbb{R}^6$ (see [Gr 3] for an account of the theory of isometric imbeddings of Riemannian manifolds). Unfortunately, the existence of isometric imbeddings $V \to \mathbb{R}^q$ has no application to the actual study of the Riemannian geometry of

V.

(a') The above example obviously generalizes to smooth submanifolds V in an arbitrary Riemannian manifold W. Namely, the restriction of the length functional on curves in W to those in V induces a Riemannian structure in V from W.

(b) Take the unit sphere $S^{n-1} \subset \mathbb{R}^n$ with the induced Riemannian metric (from the Euclidean structure in $\mathbb{R}^n$) and let P^{n-1} denote the projective space which is obtained from S^{n-1} by identifying the pairs of points s and $-s$ in S^{n-1}. Since the involution $s \to -s$ is an isometry on S^{n-1}, one obviously has a unique Riemannian metric on P^{n-1} for which the quotient map $S^{n-1} \to P^{n-1}$ is locally an isometry. Observe that P^{n-1} (unlike S^{n-1}) admits no isometric C^2-realization in $\mathbb{R}^n$. (In fact, the space P^{n-1} for $n \geq 3$ admits no isometric C^2-imbedding to $\mathbb{R}^{n+2}$, but P^2, for instance, can be isometrically C^∞-embedded into $\mathbb{R}^5$).

(b') Take linearly independent vectors $X_1, \ldots, X_n$ in $\mathbb{R}^n$ and let $\mathcal{L} \approx \mathbb{Z}^n$ be the free Abelian group (lattice) generated by these vectors. Then the quotient space, called a *flat torus*, $\mathbb{T}^n = \mathbb{R}^n/\mathcal{L}$ inherits a Riemannian metric from the Euclidean one in $\mathbb{R}^n$ for which the quotient map $\mathbb{R}^n \to \mathbb{T}^n$ is locally an isometry. The space of (isometry classes of) flat tori is (for a natural topology) a fairly complicated locally compact space of dimension $\frac{n(n+1)}{2}$. No flat torus admits an isometric imbedding into $\mathbb{R}^m$ for $m \leq 2n - 1$ but many are embeddible into $\mathbb{R}^{2n}$ (e.g. 2-tori go into $\mathbb{R}^4$ and *split* n-tori which are *isometric* products of circles obviously go into $\mathbb{R}^{2n}$ for all n).

(c) One usually defines a Riemannian structure on a smooth manifold V as a family (or a field) of Euclidean structures in the tangent spaces $T_v(V)$ which is an assignment to each $v \in V$ of a positive quadratic form g_v on $T_v(V)$. For example, if V is a smooth submanifold in $\mathbb{R}^q$, then each tangent space $T_v(V)$ is realized by the n-plane which is the n-dimensional affine subspace in $\mathbb{R}^q$ (geometrically) tangent to V at v. Each $T_v(V)$ can be brought to the origin in $\mathbb{R}^q$ by the parallel translation $T_v(V) \to T_v(V) - v$. By restricting the Euclidean form in $\mathbb{R}^q$ to this (now linear) subspace $T_v(V) - v$ in $\mathbb{R}^q$ one gets a particular field of forms in $T_v(V)$ which gives the induced Riemannian structure in V described above in the language of length.

(c') Let $X_i = X_i(v)$, $i = 1, \ldots, n$, be linearly independent tangent vector fields on an n-dimensional manifold V. Then one may uniquely define a metric g on V by putting $g(X_i, X_j) = \delta_{ij}$, where $\delta_{ii} \equiv 1$ and $\delta_{ij} \equiv 0$ for $i \neq j$. This is, in fact, a very general procedure. For example, if V is homeomorphic to $\mathbb{R}^n$ then every Riemannian metric on V comes this way with some fields X_i. Indeed, the space $\mathbb{R}^n$ with any Riemannian metric admits, by elementary topology, a frame of n orthonormal vector fields X_i.

Another (by far more interesting) example arises from Lie groups. Namely, if V is endowed with a structure of a Lie group, then each tangent vector at the neutral element $e \in V$ uniquely extends by the left group translations to a left invariant field on V. Thus each basis of vectors $X_i \subset T_e(V)$ extends to a frame on all of V and we obtain with the above what is called a left *invariant Riemannian metric* on V. In other words, left translations extend each Euclidean structure in $T_e(V)$ to a Riemannian metric on V for which left translations are isometries. These metrics are easy to define, but not at all easy to understand. For example, if V is the multiplicative group of non-singular $(m \times m)$-matrices, then $T_e(V)$ equals the linear space $M^m = \mathbb{R}^n$, $n = m^2$ of all $(m \times m)$-matrices. The standard Euclidean structure in this $\mathbb{R}^n$ uniquely extends to a Riemannian metric in the space V of non-singular matrices. We suggest to the reader to evaluate the Riemannian distance between two matrices with given entries.

1.3. Equidistant translates of hypersurfaces in $\mathbb{R}^n$. Consider a C^2-smooth *hypersurface* V_0 (i.e. a submanifold of dimension $n-1$) in $\mathbb{R}^n$ with the Euclidean metric and let V_t, $t \in \mathbb{R}$, be a smooth one-parametric deformation of V_0, that is a C^2-map, say $F : V_0 \times \mathbb{R} \to \mathbb{R}^n$, such that the map F on $V_0 = V_0 \times 0$ equals the original embedding $V_0 \hookrightarrow \mathbb{R}^n$. The deformation is called *normal* to V_0 if the derivative $\frac{dF}{dt}(v,0) \in \mathbb{R}^n$ is normal to the tangent hyperplane $T_v(V) \subset \mathbb{R}^n$ for all $v \in V$. We want to study the deformation of the induced metric, now called g_t on the manifold $V_0 = V_0 \times t$ which is mapped into $\mathbb{R}^n$ by $v \to F(v,t)$. (The image $F(V_0 \times t)$ may not be a submanifold in $\mathbb{R}^n$, but this is rather irrelevant for our present purpose). The length $length_{g_t}(C)$ of each curve $C \subset V_0$ by definition of g_t is the Euclidean length of the curve $c \to F(c,t)$. This is equivalent to saying that $g_t(X,X)$ for each vector $X \in T_v(V_0)$ equals the squared Euclidean length of the image of X under the differential of the map F on $V_0 = V_0 \times t \subset V_0 \times \mathbb{R}$. Here each $T_v(V_0) \subset \mathbb{R}^q$ is viewed as a vector space where the point $v \in T_v(V_0)$ is taken for the origin.

PROPOSITION-DEFINITION (Gauss). *There exists a unique continuous map which assigns to each $v \in V_0$ and each vector $\nu = \nu_v$ normal to V_0 at v a symmetric linear operator $A_\nu : T_v(V_0) \to T_v(V_0)$ with the following property. If F is an arbitrary deformation for which $\frac{dF}{dt}(v,0) = \nu$, then $g'_{t=0}(X,Y) = 2g_0(A_\nu X, Y)$, where X and Y are arbitrary fixed tangent vectors in $T_v(V_0)$ and g'_t stands for the derivative $\frac{dg_t}{dt}$. Furthermore, $A_{\alpha\nu} = \alpha A_\nu$ for all $\alpha \in \mathbb{R}$.*

PROOF: Identify a small neighbourhood of a fixed point in V_0 with a domain $U \subset \mathbb{R}^{n-1}$, let $u_1, \ldots, u_{n-1}$ be the Euclidean coordinates in U and let $\partial_i = \frac{\partial}{\partial u_i}$ be the corresponding derivations (fields) in U. Then the scalar products $g_t(\partial_i, \partial_j)$ by the definition of the induced metric are

$$g_t(\partial_i, \partial_j) = < \partial_i F, \partial_j F > ,$$

where $< , >$ is the Euclidean scalar product in $\mathbb{R}^n$. Since $< \nu, \partial_i F > = 0$ for $t = 0$, we have (using $\frac{dF}{dt}(v,t) = \nu$ for $t = 0$ and operating with ∂_j)

$$< \frac{d}{dt}\partial_j F, \partial_i F > = - < \nu, \partial_i \partial_j F > \text{ at } t = 0$$

and so

$$\begin{aligned} g'_{t=0}(\partial_i, \partial_j) &= \frac{d}{dt} < \partial_i F, \partial_j F > \\ &= -2 < \nu, \partial_i \partial_j F > . \end{aligned}$$

Thus the operator A_ν is uniquely defined by $g_0(A_\nu\partial_i, \partial_j) = - < \nu, \partial_i\partial_j F >$ at $t = 0$.

□

REMARK. Fix a point $v_0 \in V_0$, let ν_0 be a unit normal to V_0 at v_0 and let $\mathbb{R}^1 \subset \mathbb{R}^n$ be the 1-dimensional subspace parallel to ν_0. Take the orthogonal hyperplane $\mathbb{R}^{n-1} \approx L \subset \mathbb{R}^n$ which is parallel to $T_{v_0}(V_0) \subset \mathbb{R}^n$ and observe that the orthogonal projection $P : V_0 \to L$ *diffeomorphically* sends a small neighbourhood $U_0 \subset V_0$ of v_0 into a domain $U \subset L$. Denote by P_1 the orthogonal projection $\mathbb{R}^n \to \mathbb{R}^1$ and observe that $U_0 \subset \mathbb{R}^n$ equals the *graph* of the function $f = P_1 P^{-1} : U \to \mathbb{R}^1$. Furthermore,

$$< \nu_0, \partial_i\partial_j F >= \partial_i\partial_j f \ ,$$

for the Euclidean coordinates u_i in this U and, hence, $-\frac{1}{2}g'_{v_0}$ equals the second differential $d^2 f$ of f at $u_0 = P(v_0) \in U$.

Next consider a hypersurface $\overline{V}_0 \subset \mathbb{R}^n$ which is an *inside tangent* (relative to ν_0) to V_0 at v_0. That is $\overline{V}_0$ contains v_0, the tangent hyperplane $T_{v_0}(\overline{V}_0) \subset \mathbb{R}^n$ equals $T_{v_0}(V_0)$ and the corresponding function $\overline{f}$ satisfies $\overline{f} \geq f$ near u_0. Then $d^2\overline{f} \geq d^2 f$ and we conclude to the following (simple but useful) inequality

$$\overline{g}'_{v_0} \leq g'_{v_0} \ .$$

See the first chapter in [Mi] for additional information.

Next we turn to the following *normal geodesic* deformation $F : V_0 \times \mathbb{R} \to \mathbb{R}^n$ which isometrically sends each line $R = v \times \mathbb{R} \subset V_0 \times \mathbb{R}$ onto the straight line in $\mathbb{R}^n$ normal to V_0 at v. Such a map exists if and only if V_0 is *normally orientable* in $\mathbb{R}^n$ (unlike the Möbius band in $\mathbb{R}^3$) which is always the case for small neighbourhoods of points in V_0. We want to study the second derivative g''_t for normal geodesic maps F. We denote by $A(t)$ the operators A_ν assigned to the hypersurface $F(V_0 \times t)$. Again, our consideration is local near a fixed point $v \in V_0$ and we only allow those values t for which the map F diffeomorphically sends $U' \times t$ for a small U' around v onto a smooth hypersurface in $\mathbb{R}^n$. This is the case (by the implicit function theorem) if and only if the differential of F is injective on the tangent space $T_{v,t}(V_0, t)$. The following theorem shows this property to fail exactly for those t which equal the reciprocals of the eigenvalues of the operator $A(0)$ at v and which are called the *principal curvatures* of V_0 at v relative to the unit normal $\nu = \frac{dF}{dt}(v, 0)$.

THEOREM (Gauss-Weil). *The derivative of $A(t)$ in t satisfies*

$$A'(t) = -A^2(t) \ ,$$

where A^2 is the ordinary square of the operator A.

PROOF. Since $< \frac{dF}{dt}, \partial_j F >= 0$, we have

$$< \frac{d}{dt}\partial_i F, \partial_j F >=< \partial_i F, \frac{d}{dt}\partial_j F >= \frac{1}{2}\frac{d}{dt} < \partial_i F, \partial_j F >= \frac{1}{2}g'_t(\partial_i, \partial_j) = g_t(A(t)\partial_i, \partial_j) \ .$$

Hence,

$$g_t(A^2(t)\partial_i, \partial_j) = g_t(A(t)\partial_i, A(t)\partial_j) = < \frac{d}{dt}\partial_i F, \frac{d}{dt}\partial_j F > .$$

Since $\frac{d^2F}{dt^2} = 0$, we obtain

$$2 < \frac{d}{dt}\partial_i F, \frac{d}{dt}\partial_j F > = \frac{d^2}{dt^2} < \partial_i F, \partial_j F > = 2\frac{d}{dt} g_t(A(t)\partial_i, \partial_j)$$
$$= 2g_t'(A(t)\partial_i, \partial_j) + 2g_t(A'(t)\partial_i, \partial_j) = 4g_t(A^2(t)\partial_i, \partial_j) + 2g_t(A'(t)\partial_i, \partial_j) ,$$

which equals, by the above, $2g_t(A^2(t)\partial_i, \partial_j)$.

□

EXAMPLES. (a) Let V_0 be the unit sphere $S^{n-1} \subset \mathbb{R}^n$ and let $F(s,t) = s(1-t)$ be the geodesic deformation corresponding to the interior normals. Then, clearly, $g_t' = -2(1-t)g_0$ and therefore $A(t) = -(1-t)^{-1}Id$. Thus all principal curvatures of the sphere of radius $1-t$ equal $-(1-t)^{-1}$ for the *interior* normal direction. They become infinite at $t = 1$ as the map F collapses S^{n-1} to a single point.

(b) Let V_0 be an arbitrary *closed* C^2-hypersurface in $\mathbb{R}^n$ which bounds a compact region $V_+ \subset \mathbb{R}^n$. Consider the interior normal geodesic deformation $F : V_0 \times \mathbb{R}_+ \to \mathbb{R}^n$ and let $r(v) \in R_+ = \mathbb{R}_+ \times v$ for $v \in V_0$ be the first singular point on $\mathbb{R}_+ \times v$ where the map F fails to be a local diffeomorphism. (We know by the above that $-(r(v))^{-1}$ equals the smallest eigenvalue of $A(v,0)$). Consider the open subset $W_+ = \{v,t | v \in V_0, t < r(v)\} \subset V_0 \times \mathbb{R}$.

PROPOSITION. *The n-dimensional measure of the difference $V_+ \backslash F(W_+) \subset \mathbb{R}^n$ equals zero.*

PROOF. Take a point $x \in V_+$ and let r_x be the radius of the greatest ball B_x in V_+ around x. The boundary ∂B_x is an inside tangent to V_0 at some point $v = v(x) \in V_0$. Hence, $x = F(v, r_x)$ and $r(v) \geq r_x$. The set $F\{v, r(v)\} \subset \mathbb{R}^n$ obviously has measure zero, and so almost all points $x \in V_+$ are contained in $F(W_+)$.

□

EXERCISE. Let V_0 be a closed *convex* hypersurface whose principle curvatures relative to the exterior normal are ≤ 1. Show the normal geodesic map $F(v,t)$, where $\frac{dF}{dt}$ is the *exterior* normal field, to be *injective* on $V_0 \times (-1, \infty) \subset V_0 \times \mathbb{R}$. Show that the convex region bounded by V_0 contains a unit ball. Next consider a convex subset V_+ with smooth convex boundary $V_0 = \partial V_+$ which has the principal curvatures > 1. Define $V_t \subset V_+$ as the set of those $v \in V_+$ for which $dist(v, V_0) \geq t$. Show that each set V_t is convex. Prove, moreover, that V_t is the intersection of convex subsets with *smooth* boundaries whose principal curvatures $> (1-t)^{-1}$. In particular, the minimal non-empty V_t is a single point, say, v in V_+. Show that V_+ is contained in some ball of radius 1 in $\mathbb{R}^n$.

1.4. Normal deformation of the Riemannian volume. Each Riemannian manifold of dimension n carries a canonical measure (volume) which is uniquely defined by the following two axioms:

(1) The volume of the unit cube in $\mathbb{R}^n$ equals 1.

(2) If $f : V_1 \to V_2$ is a distance decreasing map of V_1 *onto* V_2, then $Vol\ V_2 \leq Vol\ V_1$, where $dim\ V_1 = dim\ V_2 = n$.

This is obvious since each continuous Riemannian structure g_u in $U \subset \mathbb{R}^n$ can be approximated near each point $u_0 \in U$ by the Euclidean structure g_{u_0} (which is constant g_{u_0} in U). The following computational formula for this volume is equally obvious. Let μ_0 be the Euclidean Haar measure in $\mathbb{R}^n$ and let μ_u be the Haar measure associated to the form g_u on $\mathbb{R}^n$ (for which the measure of the g_u-unit cube is one). Then the ratio $r(u) = \mu_u/\mu_0$ is a continuous function in u and $Vol\ U = \int r(u)d\mu_0$. Moreover, one defines the Jacobian of a smooth map $f : V_1 \to V_2$ between Riemannian manifolds, say $J(v)$, $v \in V_1$, as the absolute value of the determinant of the differential $D_f : T_v(V_1) \to T_{v'}(V_2)$ for $v' = f(v)$, relative to the Euclidean structures g_v in $T_v(V_1)$ and $g'_{v'}$ in $T_{v'}(V_2)$, where g_v and $g'_{v'}$ are the Riemannian structures in V_1 and V_2 respectively (which by definition are Euclidean structures in the tangent spaces). If f is a bijective map, then $Vol\ V_2 = \int_{V_1} J(v)dv$ where dv is the Riemannian measure in V_1.

EXERCISES. (a) Show every *compact* manifold V to have *finite* total volume $Vol\ V$.

(b) Show the hyperbolic plane H^2 to have $Vol\ H^2 = \infty$.

(c) Let g_0 be a Euclidean structure in $\mathbb{R}^2$ and let a continuous function $h(u)$ on $\mathbb{R}^2$ equal $(\|u\| log\|u\|)^{-2}$ outside a compact subset in $\mathbb{R}^2$ for $\|u\| = dist_{g_0}(u,0)$. Show that the Riemannian metric $g = hg_0$ on $\mathbb{R}^2$ is complete and $Vol(R^2, g) < \infty$.

(d) Show every non-compact Lie group with a left invariant metric to be complete of infinite volume.

(e) Let $SL_n\mathbb{R}$ be the group of unimodular $(n \times n)$-matrices with a left invariant metric and let $SL_n\mathbb{Z}$ be the subgroup of the matrices with integer entries. Show the quotient manifold $SL_n\mathbb{R}/SL_n\mathbb{Z}$ to be complete of *finite* volume (and noncompact for $n \geq 2$).

REMARK. The double coset space $SO_n\backslash SL_n\mathbb{R}/SL_n\mathbb{Z}$ is naturally homeomorphic to the space of flat tori $\mathbb{T}^n$ of unit volume.

Now, with the Gauss formula $g'_t(X,Y) = 2g_0(A_\nu X, Y)$ for normal deformations we immediately see the following formula for the derivative in t of the volume $Vol_t = Vol_{g_t}(U_0)$ for all domains $U_0 \subset V_0$

$$Vol'_{t=0} = \int_{U_0} trace\ A_\nu dv_0$$

for the volume element dv_0 of the metric g_0 (which is the usual volume of a submanifold in $\mathbb{R}^n$). The trace of A_ν relative to a unit normal is called the *mean curvature* of $V_0 \subset \mathbb{R}^n$ and it clearly equals the sum of principal curvatures. For example, the mean curvature of the unit sphere $S^{n-1} \subset \mathbb{R}^n$ equals $n-1$ for the exterior normal and it is $-(n-1)$ for the interior normal.

In order to apply this formula to $t \neq 0$ we express the volume element of g_t on $F(V_0 \times t) \subset \mathbb{R}^n$ by $dv_t = J(v_0,t)dv_0$ for the Jacobian of the map F. Then for all $t \in \mathbb{R}$,

$$Vol'_t = \frac{d}{dt}\int_{U_0} J(v_0,t)dv_0 = \int_{U_0}(trace\ A_\nu(v_0,t))J(v_0,t)dv_0\ .$$

This is equivalent to the relation

$$\frac{dJ(v_0,t)}{dt} = J(v_0,t)Trace\ A_\nu(v_0,t)$$

which holds on the greatest interval $a(v_0) \leq t \leq b(v_0)$ around zero where the Jacobian does not vanish. Then one exresses the above by

$$\frac{d}{dt}log\ J(v_0,t) = Trace\ A_\nu(v_0,t)\ ,$$

and the Gauss-Weil formula implies

$$\frac{d^2}{dt^2}log\ J(v_0,t) = -trace\ A^2_\nu(v_0,t) \leq -\frac{1}{n-1}(trace\ A_\nu(v_0,t))^2 = -\frac{1}{n-1}(\frac{d}{dt}log\ J(v_0,t))^2\ .$$

Now, a straightforward computation shows that

$$J(v_0,t) \leq 1 + \left(\frac{t\ Trace\ A_\nu(v_0,0)}{n-1}\right)^{n-1}\ .$$

EXERCISE. Show that $J(v_0,t) = Det(1 + tA_\nu(v_0,0))$.

1.4.A. THEOREM. (Paul Levy [Lev]) *Let V_0 be a closed hypersurface in $\mathbb{R}^n$ whose mean curvature relative to the interior normal everywhere is $\leq \overline{\mu}$ for some number $\overline{\mu} < 0$. then the region V_+ bounded by V_0 has $Vol\ V_+ \leq -\frac{n-1}{n\overline{\mu}}Vol\ V_0$. (In fact the equality holds if and only if V_0 is a round sphere of radius $-\overline{\mu}/(n-1)$.*

PROOF. Let $F:\ V_0 \times \mathbb{R}_+ \to \mathbb{R}^n$ be the interior normal geodesic map and let $W_+ = \{v_0 \in V_0,\ 0 \leq t < r(v_0)\} \subset V_0 \times \mathbb{R}_+$ be the maximal open subset on which F is locally diffeomorphic. (This is equivalent to $J(v_0,t) > 0$ for $(v_0,t) \in W_+$). Then $J(v_0,t) \leq (1+\frac{\overline{\mu}t}{n-1})^{n-1}$ which implies $r(v_0) \leq \overline{r} = -\frac{n-1}{\overline{\mu}}$ for all $(v_0,t) \in W_+$. Hence,

$$Vol\ F(W_+) \leq \int_{V_0}\int_0^{r(v_0)} J(V_0,t)dv_0dt \leq Vol\ V_0\int_0^{\overline{r}}(1+\frac{\overline{\mu}t}{n-1})^{n-1} = -\frac{n-1}{n\overline{\mu}}Vol\ V_0\ .$$

Since $meas(V_+ \backslash F(W_+)) = 0$, the proof follows.

□

1.5. Normal deformations in Riemannian manifolds. Let V be a C^∞-smooth Riemannian manifold of dimension n which is complete as a metric space. A C^1-map $f:\ \mathbb{R} \to V$ is called a *geodesic* if $dist(f(t_0), f(t)) = |t_0 - t|$ for all $t_0 \in \mathbb{R}$ and for all $t \in \mathbb{R}$ close to t_0. Let V_0 be a C^2-smooth normally oriented hypersurface in V. Then there exists a unique (with the given normal orientation) *normal geodesic map* $F: V_0 \times \mathbb{R} \to V$ for which the curve $F_{|v_0 \times \mathbb{R}}: R = v_0 \times \mathbb{R} \to V$ is a geodesic normal to V_0 at v_0 for all $v_0 \in V_0$ – (see [Mi], [G.K.M.], [C.E.]). We define as above the family g_t of the induced Riemannian metrics on $V_0 = V_0 \times t$ and we study g'_t and g''_t at $t = 0$ as earlier. First we define the operators $A_\nu = A(v_0, t)$, where the normal direction ν in V is understood as the image of the field $\frac{\partial}{\partial t}$ on $V_0 \times \mathbb{R}$ under the differential of the map F, by setting $g_t(A_\nu X, Y) = \frac{1}{2} g'_t(X, Y)$ for all pairs of tangent vectors X and Y in V_0.

PROPOSITION-DEFINITION (Riemann). *There exists a smooth map $\nu \to K_\nu$ which for each $v_0 \in V_0 \subset V$ assigns to the unit vector $\nu \in T_{v_0}(V_0)$, normal to $T_{v_0}(V_0) \subset T_{v_0}(V)$ for all $v_0 \in V$, an operator $K_\nu: T_{v_0}(V_0) \to T_{v_0}(V_0)$, called normal curvature operators, such that $\frac{d}{dt} A_\nu = -A_\nu^2 - K_\nu$ at $t = 0$ for all normal vectors. Furthermore, the operator K_ν depends only on ν (and on the Riemannian metric in V, of course) but not on V_0. That is, if V_0 and V'_0 have a common tangent space at some point v_0, that is $v_0 \in V_0 \cap V'_0$ and $T = T_{v_0}(V_0) = T_{v_0}(V'_0) \subset T_{v_0}(V)$, then the operator $K_\nu: T \to T$ defined by $K_\nu = -(\frac{d}{dt} A_\nu + A_\nu^2)$ for $V_0 \cap V'_0$, automatically satisfy the same relation for V'_0. (Observe that the operator A_ν, unlike K_ν, does depend on V'_0.)*

The proof (by a straightforward computation) can be found in any textbook on Riemannian geometry (e.g. [G.K.M.], [C.E.], where an equivalent language of Jacoby fields is employed).

EXAMPLES. (a) Let V be the round sphere S^n in $\mathbb{R}^{n+1}$ of radius R and let V_0 be a round sphere S^{n-1} in S^n whose points are all within (geodesic in S^n) distance r_0 from a fixed point $v \in V = S^n$. then, clearly, the metric g_t on $V_0 = V_0 \times t$ satisfies $g_t = \rho\, sin^2[(t + r_0)/R] g_0$ for $\rho = sin^2(r_0/R)$. Then $g'_t = (\frac{2\rho}{R} sin \frac{t+r_0}{R} cos \frac{t+r_0}{R}) g_0$ and so $A_\nu = (R^{-1} ctg \frac{t+r_0}{R}) Id$. Next, $A'_\nu = (-R^{-2} - R^{-2} ctg(\frac{t+r_0}{R})^2) Id$ and so $K_\nu = R^{-2} Id$. Thus the sphere S^n of radius R has (constant) curvature R^{-2}.

(b) <u>The exponential map.</u> There exists a unique map $T_v(V) \to V$ for each point $v \in V$, called $exp: T_v(V) \to V$, which isometrically sends each straight line ℓ in $(T_v(V), g_v) \approx \mathbb{R}^n$ through the origin onto a geodesic γ in V through v which is tangent at v to ℓ. (See [Mi], [G.K.M.], [C.E.].) It follows that each ball $\tilde{B}_\varepsilon \subset T_v(V)$ around the origin of *small* radius $\varepsilon > 0$ is diffeomorphically sent onto the ε-ball $B_\varepsilon \subset V$ around v. Hence, the boundary S_ε of B_ε for small ε is a C^∞-smooth hypersurface in V, whose interior normal geodesic map $F: S_\varepsilon \times \mathbb{R}_+ \to V$ diffeomorphically sends $S_\varepsilon \times [0, \varepsilon)$ onto $B_\varepsilon \backslash \{v\}$, while the map $F(s, \varepsilon)$ collapses S_ε to v. This implies, like in the Euclidean case, the following property of the normal geodesic map F of an arbitrary hypersurface V_0 in V.

If for a fixed point $x_0 \in V$ the function $dist(x_0, v)$, $v \in V_0$, assumes a local minimum at some point $v_0 \in V_0$ then $F(v_0, r) = x_0$ for $r = dist(x_0, v_0)$ (and for the obvious choice of the normal ν_0 at v_0) and the map F is locally diffeomorphic at (v_0, t) for $0 \le t < r$.

COROLLARY. *Let V_0 be a closed hypersurface in V which bounds a compact re-*

gion $V_+ \subset V$, let $F : V_0 \times \mathbb{R}_+ \to V$ be the interior normal geodesic map and let $W_+ = \{v_0 \in V_0,\ 0 \le t < r(t)\} \subset V_0 \times \mathbb{R}_+$ be the maximal open subset on which F is locally diffeomorphic. Then $meas_n(V_+ \backslash F(W_+)) = 0$.

REMARK. We shall need below a simple generalization of this fact to compact subsets $V_+ \subset V$ with possibly non-smooth boundary. Namely, let V_0 be the topological boundary of V_+, let $V_0' \subset V_0$ be the maximal open subset which is a C^∞-smooth hypersurface and let $\Sigma = V_0 \backslash V_0'$ be the complementary *singular part* of V_0. The singularity Σ is called *negligible* for V_+ if no compact subset $V_+' \subset V_+$ with C^∞-*smooth* $(n-1)$-dimensional boundary ever meets Σ.

EXAMPLE. Take a domain D in S^2 bounded by a smooth curve C in S^2 and let $V_+ \subset \mathbb{R}^3 \supset S^2$ be the cone over D from the origin. The boundary $V_0 = \partial V_+$ is singular at the origin unless C is the great circle and this singularity is negligible if and only if D contains no hemisphere.

Now, for V_0 with a negligible singularity we define the map F outside Σ, that is $F : V_0' \times \mathbb{R}_+ \to V$ and the image of the corresponding $W_+' \subset V_0' \times \mathbb{R}_+$ obviously covers almost all of V_+.

EXERCISES. (a) Show that the curvature of the hyperbolic plane everywhere is $-Id$.

(b) A hypersurface V_0 is called *convex* (*concave*) for a given normal orientation if the operators A_ν are positive (negative). Show that normal geodesic deformations preserve convexity if the (symmetric) operator K_ν is negative for all ν. Similarly, these deformations preserve the concavity if $K_\nu \ge 0$.

RICCI CURVATURE. Define *Ricci* ν for all unit vectors ν in $T(V)$ by *Ricci* ν = *Trace* K_ν. For example, the round n-dimensional sphere of radius R has *Ricci* $\nu = const = (n-1)R^{-2}$. The following (by now obvious but important) formula generalizes the volume deformation property from $\mathbb{R}^n$ to all V.

1.5.A. LEMMA. *The Jacobian of the map F satisfies*

$$\frac{d}{dt} \log J(v_0,t) = Trace\ A_\nu(v_0,t)\ ,$$

and

$$\frac{d^2}{dt^2} \log J(v_0,t) = -trace\ A_\nu^2(v_0,t) - Ricci\ \nu \le -\frac{1}{n-1}\left(\frac{d}{dt} \log J(v_0,t)\right)^2 - Ricci\ \nu\ .$$

Observe that the equality here holds for spheres $V_0 = S^{n-1} \subset V = S^n$.

1.5.B. THEOREM. (Paul Levy). *Let a closed submanifold V_0 with a negligible singularity in V bounds a compact region $V_+ \subset V$ such that the mean curvature (that is Trace A_ν for the interior normal) of the non-singular locus $V_0' \subseteq V_0$ everywhere is $\le \bar{\mu}$, for a given $\bar{\mu} \in \mathbb{R}$ (of any sign now) and let the Ricci curvature of V everywhere $\ge (n-1)R^{-2}$ for some $R > 0$. Let V_+^* be the ball of the (geodesic) radius r_0 in the round sphere $S^n \subset \mathbb{R}^{n+1}$ of radius R, where*

$0 \le r_0 \le \pi R$ such that $R^{-1} ctg \frac{r_0}{R} = -\bar{\mu}/(n-1)$. (The boundary V_0^* of this ball has constant mean curvature $= \bar{\mu}$). Then, $Vol\ V_+ \le \frac{(Vol\ V_+^*) Vol\ V_0}{Vol\ V_0^*}$.

PROOF. The Jacobian $J(v_0, t)$ of the interior normal map on

$$W_+' = \{v_0 \in V_0',\ 0 \le t < r(t)\} \subset V_0' \times \mathbb{R}_+$$

is majorized by the Jacobian of the corresponding map for V_0^*, that is $J(v_0, t) \le J^*(v_0^*, t)$, where the Jacobian J^* (obviously) depends only on t but not on v_0^*. This majorization is obtained by a direct computation with the above formulae. It follows that $r(v_0) \le r_0$ for all $v_0 \in V_0'$ and that

$$\begin{aligned} Vol\ V_+ &\le Vol\ F(W_+') \le \int_{V_0'} \int_0^{r(v_0)} J(v_0, t) dv_0 dt \\ &\le \frac{Vol\ V_0'}{Vol\ V_0^*} \int_{V_0^*} \int_0^{r_0} J^*(t) dv_0^* dt = \frac{Vol\ V_0'\ Vol\ V_+^*}{Vol\ V_0^*} . \end{aligned}$$

This is obvious for those who have confidence in their Riemannian geometry. A novice is invited to check all the details step by step by comparing with the Euclidean case.

EXERCISES. (a) Let V_+ be a compact region in V with possibly non-smooth boundary. Say that the mean curvature of the boundary is $\ge \bar{\mu}$ if V_+ is the intersection of a decreasing sequence of regions with smooth boundaries which have mean curvatures $\ge \bar{\mu}$. Take the ε-neighbourhood of the boundary, say $U_\varepsilon(\partial V) \subset V$ and define $Vol\ \partial V = liminf_{\varepsilon \to 0} \varepsilon^{-1} Vol(U_\varepsilon(\partial V) \cap V_+)$, where Vol on the right-hand- side denotes the n-dimensional Riemannian volume (or rather the measure) which is obviously defined for all Borel subsets in V. Generalize the above theorem to these regions V_+.

(b) Let V have $Ricci\ \nu \ge 0$ for all unit vectors $\nu \in T(V)$. Show that concentric balls B_1 and B_2 of radii R_1 and $R_2 \ge R_1$ have

$$\frac{Vol\ B_1}{Vol\ B_2} \ge \left(\frac{R_1}{R_2}\right)^n .$$

(b') Let $Ricci\ \nu \ge (n-1)R^{-1}$, for $R > 0$, and let B_1^* and B_2^* be the balls of radii R_1 and R_2 in the sphere S_R^n of radius R. Show that

$$\frac{Vol\ B_1}{Vol\ B_2} \le \frac{Vol\ B_1^*}{Vol\ B_2^*} \tag{$*$}$$

and prove as a corollary that

$$Vol\ V \le Vol\ S_R^n \quad \text{and} \quad diam\ V \le diam\ S_R^n = \pi R .$$

(c) Generalize the above $(*)$ to the case $Ricci\ \nu \ge (n-1)R^{-2}$ for $R < 0$, by replacing the sphere S_R by a pertinent hyperbolic space (or, alternatively, by writing explicit formulae in place of $Vol\ B_1^* / Vol\ B_2^*$).

(d) Show all balls in a complete simply connected manifold of *negative curvature* (that is, the operators K_ν are negative) to be convex (i.e. to have the convex boundary for the exterior normal).

(e) Let a complete manifold V have $K_\nu \geq R^{-2}Id$ for some $R > 0$. Show that the complement to each ball in V of radius $\geq \pi R$ is a convex subset in V(i.e. it is an intersection of regions with smooth convex boundaries).

2. ISOPERIMETRIC INEQUALITIES

2.1. Inequalities for Banach spaces. The *classical isoperimetric inequality* in $\mathbb{R}^n$ claims that among all domains $V_+ \subset \mathbb{R}^n$ of a given volume the round ball B^n has the minimal $(n-1)$-dimensional volume of the boundary,

$$Vol\ V_+ \leq C_n(Vol\ \partial V_+)^{n/n-1} \tag{1}$$

where $C_n = Vol\ B^n/(Vol\ S^{n-1})^{n/n-1}$ for the unit sphere $S^{n-1} = \partial B^n$. This can be equally expressed with the characteristic function f_+ of V_+ (which is 1 on V_+ and 0 outside) by *Sobolev's inequality* which relates the L_p-norm of functions on $\mathbb{R}^n$ for $p = \frac{n}{n-1}$ to the L_1-norm of the differentials df,

$$\left(\int_{\mathbb{R}^n} |f|^{n/n-1}\right)^{(n-1)/n} \leq C_n^{(n-1)/n} \int_{R^n} \|df\| du \tag{2}$$

for all functions f on $\mathbb{R}^n$ with a compact support and where du is the (Euclidean) Haar measure in $\mathbb{R}^n$. If f is not differentiable then df is understood in the sense of distributions. For example,

$$\int_{\mathbb{R}^n} \|df_+\| du = Vol\ \partial V_+ \text{ for compact domains } V_+ \subset \mathbb{R}^n \text{ with } C^1 - \text{boundary}\ .$$

This can be seen with the approximation $f_\varepsilon \to f_+$, $\varepsilon \to 0$, where $f_\varepsilon(u) = f_+(u)$ for $u \in V_+$ and $f_\varepsilon(u) = max(0, 1 - dist(u, V_+))$ outside V_+. Hence the inequality (2) for f_+ reduces to (1). On the other hand, the inequality (2) can be obtained by applying (1) to the regions $V_t = \{u \in \mathbb{R}^n |\ |f(u)| \leq t\}$, $t \geq 0$, and by a (clever) integration in t (see [Maz], [Bu.Ma.]).

The inequality (2) is a member of a large family of inequalities between various L_p norms of f and df. For example, the *Poincare inequality* for functions f in a compact domain $U \subset \mathbb{R}^n$ with a smooth boundary claims

$$\int_U |f|^2 du \leq C(U) \int_U \|df\|^2 du\ , \tag{3}$$

where we assume $\int_U f\ du = 0$ and where $C(U) = (\lambda_1(U))^{-1}$ for the eigenvalue λ_1 of the Laplace operator $-\Delta$ on U (with Neumann's boundary condition). The non-trivial content of

(3) is the implied inequality $C(U) < \infty$ which amounts to $\lambda_1(U) > 0$. In fact, the inequality (3) can be obtained like (2) from an appropriate isoperimetric inequality for U. Namely, consider hypersurfaces $V_0 \subset U$ which have $\partial V_0 \subset \partial U$ and which divide U into two regions U_+ and U_- in U. Let

$$I(V_0) = min(Vol\ U_+, Vol\ U_-)/Vol V_0$$

and let $Is(U)$ be the supremum of $I(V_0)$ over all $V_0 \subset U$. Then *Cheeger's inequality* (see [Bus]) claims

$$\lambda_1(U) \geq (2\ Is(U))^{-2}\ . \tag{4}$$

which means

$$\int_U f^2(u)du \leq 2\ Is(U) \int_U \|df\|^2 du \tag{5}$$

for all functions f in U which have $\int_U f(u)du = 0$. Thus (5) is proven like (2) by applying the isoperimetric inequality to the levels of the function f (see [Ch]).

The inequalities (1) and (2) were generalized by Brunn back in 1888 to an arbitrary n-dimensional normed (Banach) space $X = (X, \|\ \|)$. Namely, let the Haar measure dx in X be normalized to have the unit ball $B = \{x \in X \mid \|x\| \leq 1\}$ of volume one (which disagrees with the Euclidean convention but has an advantage of simpler formulae) and let $\|\ \|^*$ denote the norm in the dual space X^*.

THEOREM. (Brunn [Br]). *An arbitrary C^1-function f on X with a compact support satisfies*

$$\left(\int_X |f(x)|^{n/(n-1)}\right)^{(n-1)/n} dx \leq n^{-1} \int_X \|df(x)\|^* dx\ . \tag{6}$$

PROOF. (Knothe [Kn]) Fix a linear coordinate system $x_1, \ldots, x_n$ in X such that $dx_1, dx_2, \ldots, dx_n = dx$ and let $\mu(x)$ be a continuous function on X with a compact support $S \subset X$, whose interior is denoted by $S^0 \subset S$.

LEMMA. *There exists a C^1-map Y of S^0 into the cube $C = \{0 < x_i < 1\} \subset X$ with the following two properties:*

(1) The map Y is triangular: the i-th coordinate function of Y depends on $x_1, \ldots, x_i$ only. That is

$$Y = (y_1(x_1), y_2(x_1, x_2), \ldots, y_n(x_1, \ldots, x_n))\ .$$

(2) The partial derivatives $\frac{\partial y_i}{\partial x_i}$ are non-negative on S^0 and the Jacobian

$$J(x) = Det(\frac{\partial y_i}{\partial x_j}) = \prod_{i=1}^{n} \frac{\partial y_i}{\partial x_i} \quad satisfies \quad J(x) = \mu(x) / \int_S \mu(x)dx, \quad for\ all \quad x \in S^0\ .$$

PROOF. For $s = (x_1(s), \ldots, x_n(s)) \in S^0$ set

$$A_i(s) = \{x = (x_1, \ldots, x_n) \mid x_j = x_j(s) \text{ for } j < i \text{ and } x_i \leq x_i(s)\}$$

and

$$B_i(s) = \{x = (x_1, \ldots, x_n) \mid x_j = x_j(s) \ \text{ for } \ j < i\} .$$

Then the map $Y(s) = \{y_i(s)\}$ with

$$y_i(s) = \int_{A_i(s)} \mu(s) dx_i, \ldots, dx_n / \int_{B_i(s)} \mu(s) dx_i, \ldots, dx_n \ , \ i = 1, \ldots, n$$

clearly satisfies (1) and (2).

□

REMARK. The above formula equally applies to the characteristic function of an arbitrary open convex subset U in X which gives a *triangular C^1-diffeomorphism of U onto C* whose Jacobian identically equals $(Vol\ U)^{-1}$. We are especially interested in the inverse of this diffeomorphism in case U is the open unit ball $B = \{\|x\| < 1\} \subset X$. This inverse map, called $Y^* : C \to B$, clearly is triangular with the Jacobian $\equiv 1$.

We now compose the above map $Y = Y_\mu$ with Y^* and thus obtain a C^1-map

$$(z_1, \ldots, z_n) = Z = Y^* \circ Y : S^0 \to B$$

which satisfies the above properties (1) and (2) as well as Y. This Z obviously has $\|Z(x)\| < 1$ for all $x \in S^0$ and the divergence $div\ Z(x) = \Sigma_{i=1}^n \frac{\partial Z_i}{\partial x_i}(x)$ satisfies by the geometric-arithmetic mean inequality,

$$(div\ Z(x))^n \geq n^n \mu(x) / \int_S \mu(x) dx, \quad \text{for all} x \in S^0 .$$

Finally, we take $\mu(x) = |f(x)|^{n/n-1}$ and get

$$0 = \int_S div(|f(x)| Z(x)) dx = \int_S |f(x)| div\ Z(x) dx + \int_S < d|f(x)|,\ Z(x) > dx \ ,$$

for the canonical bilinear pairing $<\ ,\ >$ between X^* and X. Therefore,

$$\int |f(x)|^{n/n-1} dx = \int |f(x)| \mu(x)^{1/n} dx \leq n^{-1} \left(\int |f(x)| div\ Z(x) dx\right) \left(\int \mu(x) dx\right)^{1/n}$$
$$\leq n^{-1} \left(\int \|df(x)\| dx\right) \left(\int \mu(x) dx\right)^{1/n} .$$

Since the inequality (6) is homogeneous, we may normalize to $\int \mu(x) dx = 1$, and then (6) follows from the above.

□

REMARKS. (a) The inequality (6) generalizes with an obvious approximation argument to all functions f with (generalized) derivatives in L_1 which, for $X = \mathbb{R}^n$, amounts to Sobolev's inequality (2).

(b) The above proof shows that the equality in (6) holds if and only if f is a scalar multiple of the characteristic function of a metric ball in X. In fact, the proof gives an integral formula for the *isoperimetric deficiency*

$$\|f(x)\|_{L_{n/n-1}} - n^{-1}\|df\|^*_{L_1} .$$

(c) The inequality (6) (and its proof as well) obviously generalizes to the spaces X with *non-symmetric* norms. This generalization (in a slightly different but equivalent form) is called the *Brunn-Minkowski inequality* (see [Had]).

(d) Brunn's inequality generalizes, up to a certain extent, to *differential forms* f on X of degree > 1, as was discovered (in the dual isoperimetric language by Federer and Fleming (see [F.F], [Gr2]).

2.2. Levy's Inequality. Let V be a closed n-dimensional Riemannian C^∞-manifold whose Ricci curvature is everywhere $\geq (n-1)R^{-2}$, $R > 0$, (which is the Ricci curvature of the round sphere $S^n \subset \mathbb{R}^{n+1}$ of radius R). Let $V_+ \subset V$ be a compact region with smooth boundary $V_0 = \partial V_+$ and let V_+^* be a round ball in the sphere S^n of radius R in $\mathbb{R}^{n+1}$ such that $Vol\ V_+^*/Vol\ S^n = Vol\ V_+/Vol\ V$.

THEOREM. (Levy [Lev]). *The $(n-1)$-dimensional volume of V_0 is related to that of the sphere $V_0^* = \partial V_+^*$ by the inequality*

$$\frac{Vol\ V_0}{Vol\ V} \geq \frac{Vol\ V_0^*}{Vol\ S^n} .$$

PROOF. Consider the functional $Vol\ \partial\Omega_+$ on all domains $\Omega_+ \subset V$ with a fixed n-dimensional volume $Vol\ \Omega_+ = Vol\ V_+$. Then, by the global calculus of variations, there exists an *extremal* domain, say $\overline{\Omega}_+$ in V for which the $(n-1)$-volume $\partial\overline{\Omega}_+$ is the least possible. Unfortunately, the boundary $\partial\overline{\Omega}_+$ is not necessarily smooth. However, a deep theorem of F. Almgren [Al] claims the singularity to be negligible both from inside and outside $\overline{\Omega}_+$. (Notice that $\partial\overline{\Omega}_+$ is smooth for $n \leq 7$, see [Law].) Furthermore, the non-singular locus $\overline{\Omega}'_0 \subset \overline{\Omega}_0 = \partial\overline{\Omega}_+$ of this Ω has *constant* mean curvature (this is obvious since the mean curvature equals the normal derivative of the $(n-1)$-dimensional volume of $\overline{\Omega}'_0$). If this curvature, called $\overline{\mu}$, does not exceed that of V_0^*, then the proof follows from 1.5.B. Otherwise, we go to the complement $\overline{\Omega}_- = V\backslash\overline{\Omega}_+$, in which (interior) direction the mean curvature of $\overline{\Omega}'_0$ equals $-\overline{\mu}$ which is necessarily less than the mean curvature of V_0^* in the direction of the ball $V_-^* = S^n\backslash V_+^*$. Then 1.5.B applies to $\overline{\Omega}^*_-$ and the proof is concluded.

□

REMARKS AND COROLLARIES. (a) If $V = S^n$, then Levy's inequality amounts to the classical isoperimetric inequality on S^n:

Among all domains in S^n with a fixed volume the minimal volume of the boundary is assumed by a round ball.

A similar inequality holds true in the n-dimensional hyperbolic space H^n (see [Schm]), but for no space except $\mathbb{R}^n$, S^n and H^n one knows the exact solution of the isoperimetric problem.

(b) Let $V_\varepsilon \subset V$ denote the ε-neighbourhood of V_0 in V. Then an obvious integration in ε shows that

$$\frac{Vol\, V_\varepsilon}{Vol\, V} \geq \frac{Vol\, V_\varepsilon^*}{Vol\, S^n} .$$

(c) Levy's inequality (and his proof as well) generalizes to all Riemannian manifold with a given (possibly negative) lower bound on the Ricci curvature (see [Gr1]). This leads to sharp estimates on the eigenvalues $\lambda_1 \leq \lambda_2 \leq \ldots$ of the Laplace operator on V (see [B.G.M.], [Ber]).

(d) In order to apply Levy's inequality to a specific manifold V one needs some information on the Ricci curvature. In fact, the Ricci curvature is an easily computable invariant.

EXAMPLES. (1) Let $V^n \subset \mathbb{R}^{n+1}$ be a smooth hypersurface whose principal curvatures $x_1, \ldots, x_n$ at some point v_0 satisfy

$$x_1(x_2 + \ldots + x_n) \geq \alpha, \quad x_2(x_1 + x_3 + \ldots x_n) \geq \alpha, \ldots, x_n(x_1 + \ldots + x_{n-1}) \geq \alpha .$$

Then $Ricci(\nu, \nu) \geq \alpha$ for all $\nu \in T_{v_0}(V)$.

This follows from the famous *theorema egregium* of Gauss which expresses the intrinsic curvature of V in terms of the principal curvatures (see [G.K.M] or prove it yourself).

(2) Let Riemannian manifolds $V_1, \ldots, V_k$ have $Ricci(V_j) \geq \alpha$, $j = 1, \ldots, k$. Then the *Riemannian product* $V = V_1 \times \ldots \times V_k$ also have $Ricci(V) \geq \alpha$. This is immediate from the definition of the Riemannian structure g in V, which is $g = g_1 \oplus \ldots \oplus g_k$.

(3) Let V be the orthogonal group $0(n)$ and let g be the (natural) left invariant metric on $0(n)$ which is invariant under conjugations (which is equivalent to being right invariant as well as left invariant) and such that the circle consisting of the rotations around a fixed subspace $\mathbb{R}^{n-2} \subset \mathbb{R}^n$ has length 2π. (With these conditions g obviously is unique)then $Ricci(0(n), g) \geq \frac{n}{4}$ everywhere (see [C.E.] for an explicit computation of the curvature of Lie groups).

APPENDIX II

GAUSSIAN AND RADEMACHER AVERAGES

II.1. THEOREM: *For all $C < \infty$ and $2 \leq q < \infty$ there exists a constant $K = K(C,q)$ such that if $\beta_q(X) \leq C$ then for all n and $x_1, \ldots, x_n \in X$*

$$\left\| \sum_{i=1}^{n} g_i x_i \right\|_{L_q(X)} \leq K \left\| \sum_{i=1}^{n} r_i x_i \right\|_{L_q(X)} .$$

In particular, $C_q(X) \leq K\beta_q(X)$ ($(g_i)_{i=1}^n$ are independent symmetric gaussian variables normalized in L_2, $(r_i)_{i=1}^n$ are the Rademacher functions).

Given a 1-unconditional basis $z_1, z_2, \ldots$ in some Banach space X and a function $f : \mathbb{R} \to \mathbb{R}$, we may view f as acting on the finitely supported elements in X by applying f to each of the coefficients

$$f(\Sigma a_n z_n) = \Sigma f(a_n) z_n .$$

II.2. DEFINITION: *The p-convexity constant A_p (resp. q-concavity constant B_q)* of the basis (z_i) is the smallest A (resp. B) satisfying

$$\|(\Sigma|x_n|^p)^{1/p}\| \leq A(\Sigma\|x_n\|^p)^{1/p}$$
$$(\text{resp.} \quad (\Sigma\|x_n\|^q)^{1/q} \leq B\|(\Sigma|x_n|^q)^{1/q}\|)$$

for all finite sequences (x_n) of finitely supported elements in X.

II.3. DEFINITION: *The upper p-estimate constant a_p* (resp. *lower q-estimate constant b_q*) of the basis (z_i) is the smallest a (resp. b) satisfying

$$\|\Sigma x_n\| \leq a(\Sigma\|x_n\|^p)^{1/p}$$
$$(\text{resp.} \quad (\Sigma\|x_n\|^q)^{1/q} \leq b\|\Sigma x_n\|)$$

for all finite sequences (x_n) of finitely and *disjointly* supported (with respect to (z_i)) elements in X.

Clearly, $a_p \leq A_p$, $b_p \leq B_p$. Also for $1 \leq p \leq \infty$, $\frac{1}{p} + \frac{1}{q} = 1$,

$$A_p(z_i) = B_q(z_i^*) \quad , \quad A_p(z_i^*) = B_q(z_i) ,$$

$$a_p(z_i) = b_q(z_i^*) \quad , \quad a_p(z_i^*) = b_q(z_i)$$

where (z_i^*) are the biorthogonal functionals to (z_i) (see [L.T2]). Note also that $a_p \leq \alpha_p(X)$, $b_q \leq \beta_q(X)$. (α_p, β_q are the gaussian type and cotype constants).

II.4. PROPOSITION: *For all $1 < r < q < \infty$ there exists a constant $K(r,q)$ such that*

$$B_q \leq K(r,q) b_r \quad \textit{for} \quad 1 < r < q < \infty$$

and

$$A_p \leq K(s,p) a_s \quad \textit{for} \quad 1 < p < s < \infty .$$

In particular

$$B_q \leq K(r,q)\beta_r(X) \quad \textit{for} \quad 1 < r < q < \infty$$

and

$$A_p \leq K(s,p)\alpha_s(X) \quad \textit{for} \quad 1 < p < s < \infty .$$

PROOF: Let $1 < p < s < \infty$. We shall show $A_p \leq K(s,p)a_s$, the result for B_q, b_r follows by duality.

We need the following lemma, the proof of which we delay until after the proof of the proposition.

II.5. LEMMA: *Let $1 < p < s < \infty$. There exist a sequence of independent random variables (f_i) on some probability space $(\Omega, \mathcal{F}, \mu)$ and a constant $K = K(s,p)$ such that*

$$K^{-1} \int_\Omega \left(\sum_{i=1}^n |a_i f_i|^s \right)^{1/s} d\mu \leq \left(\sum_{i=1}^n |a_i|^p \right)^{1/p} = \int \max_{1 \leq i \leq n} |a_i f_i| d\mu$$

for all finite sequences $a_1, \ldots, a_n$.

Assume the lemma and let $x_1, \ldots, x_n$ be finitely supported elements in X

$$x_i = \sum_j a_{ij} z_j .$$

For each $\omega \in \Omega$ we define the function $\max_{1 \leq i \leq n} |x_i f_i(\omega)|$ coordinatewise:

$$\max_{1 \leq i \leq n} |x_i f_i(\omega)| = \sum_j \max_i |a_{ij} f_i(\omega)| z_j .$$

Note that for each ω

$$\max_{1 \leq i \leq n} |x_i f_i(\omega)| = \sum_{i=1}^n y_i(\omega)$$

where the $y_i(\omega)$-s are disjointly supported elements of X and $|y_i(\omega)| \leq |x_i f_i(\omega)|$ in the order induced by (z_i). It follows from the lemma that

$$(\Sigma |x_i|^p)^{1/p} \leq \int \max_{1 \leq i \leq n} |x_i f_i(\omega)| = \int \sum_{i=1}^n y_i(\omega) .$$

By the 1-unconditionality of (z_i)

$$\|(\Sigma|x_i|^p)^{1/p}\| \le \| \int \sum_{i=1}^{n} y_i(\omega)\| \le \int \| \sum_{i=1}^{n} y_i(\omega)\| \le a_s \int (\Sigma\|y_i(\omega)\|^s)^{1/s}$$

$$\le a_s \int (\Sigma\|x_i\|^s |f_i(\omega)|^s)^{1/s} \le K\ a_s(\Sigma\|x_i\|^p)^{1/p}$$

and

$$A_p \le K\ a_s\ .$$

□

II.6. PROOF OF LEMMA II.5: Let $(f_i(\omega))_{i=1}^{\infty}$ be independent random variables with

$$\mu(f_i(\omega) > \lambda) = 1 - e^{-c/\lambda^p} \quad for\ all \quad \lambda > 0$$

where c is such that $\int f_i d\mu = 1$. Then

$$\mu(\max_{1\le i\le n} |a_i f_i| \le \lambda) = \prod_{i=1}^{n} \mu(a_i f_i \le \lambda) = exp\frac{-c\Sigma|a_i|^p}{\lambda}\ .$$

We conclude that, if $\Sigma_{i=1}^{n}|a_i|^p = 1$, then $\max_{1\le i\le n}|a_i f_i|$ has the same distribution as f_1 and in particular $\int \max_{1\le i\le n}|a_i f_i| d\mu = 1$. This proves the right hand side equality. To prove the left hand side inequality let $p < s < \infty$ and let $(g_i)_{i=1}^{\infty}$ be independent random variables such that

$$\mu(g_i > \lambda) = 1 - e^{-d/\lambda^s} \quad \text{for all} \quad \lambda > 0$$

where d is such that $\int g_i d\mu = 1$. Then, by the first part of the proof

$$\int \left(\sum_{i=1}^{n} |a_i f_i|^s\right)^{1/s} = \int\int \max_{1\le i\le n} |a_i f_i g_i|$$

$$= \int \left(\sum_{i=1}^{n} |a_i g_i|^p\right)^{1/p} \le \left(\sum_{i=1}^{n} |a_i|^p \int |g_i|^p\right)^{1/p}$$

$$= K \left(\sum_{i=1}^{n} |a_i|^p\right)^{1/p}$$

with

$$K = \left(\int |g_i|^p\right)^{1/p} < \infty\ .$$

□

II.7. Before proceeding to the proof of the Theorem we need another inequality which is a generalization due to Maurey of Khinchine's inequality.

THEOREM: *for all C and q, $1 \le C$, $q < \infty$, there exists a constant $M = M(C,q)$ such that, if (x_i) is a 1-unconditional basis in some Banach space X with q concavity constant $\le C$, then*

$$\frac{1}{\sqrt{2}}\|(\sum_i |x_i|^2)^{1/2}\| \le \int_0^1 \|\sum_i r_i(t)x_i\| \le (\int_0^1 \|\sum r_i(t)x_i\|^q)^{1/q} \le M\|(\sum_i |x_i|^2)^{1/2}\|$$

for all (x_i) in X.

PROOF: The left hand side inequality follows immediately from the triangle inequality and Khinchine's inequality in L_1, the concavity assumption is not needed here. For the right hand side inequality:

$$\begin{aligned}(\int \|\sum r_i(t)x_i\|^q)^{1/q} &\le C\|(\int |\sum r_i(t)x_i|^q)^{1/q}\| \\ &\le CK_q\|(\sum |x_i|^2)^{1/2}\|\end{aligned}$$

where K_q is Khinchine's constant in L_q. □

We turn now to the

II.8. PROOF OF THEOREM II.1: Let $\tilde{x}_i = r_i(t)x_i$ in $L_q(X)$. Since $\|\Sigma r_i(s)x_i\|_{L_q(X)} = \|\Sigma r_i(s)\tilde{x}_i\|_{L_q(L_q(X))}$ and $\|\Sigma g_i(s)x_i\|_{L_q(X)} = \|\Sigma g_i(s)\tilde{x}_i\|_{L_q(L_q(X))}$ and since $L_q(X)$ has the same cotype q constant as X, we may assume we are dealing with $(\tilde{x}_i)$ instead of (x_i). In particular, without loss of generality, (x_i) is 1-unconditional. Fix $r > q$, then, by Proposition II.4, (x_i) is r-concave with r-concavity constant depending on q, r and C only.

Let $(r_{ij})_{i=1}^{n}\,_{j=1}^{\infty}$ be a sequence of independent Rademacher functions. By Theorem II.7, for all $m \in \mathbb{N}$,

$$\begin{aligned}(\int_0^1 \|\sum_{i=1}^n\sum_{j=1}^m r_{ij}(t)\frac{1}{\sqrt{m}}x_i\|^q)^{1/q} &\le (\int_0^1 \|\sum_{i=1}^n\sum_{j=1}^m r_{ij}(t)\frac{1}{\sqrt{m}}x_i\|^r)^{1/r} \\ \le M\|(\sum_{i=1}^n |x_i|^r)^{1/r}\| &\le M\sqrt{2}(\int \|\sum_{i=1}^n r_i(t)x_i\|^q)^{1/q}\,.\end{aligned}$$

By the central limit theorem, the sequence

$$\left(\frac{1}{\sqrt{m}}\sum_{j=1}^m r_{ij}(t)\right)_{i=1}^n$$

tends in distribution to $(g_i)_{i=1}^n$, a sequence of independent, mean zero, gaussian random variables normalized in L_2. Thus we get

$$\left(\int_0^1 \|\sum_{i=1}^n g_i(t)x_i\|^q\right)^{1/q} \le M\sqrt{2}\left(\int \|\sum_{i=1}^n r_i(t)x_i\|^q\right)^{1/q}.$$

□

APPENDIX III
KAHANE'S INEQUALITY

We bring here a proof, essentially due to C. Borell, of Kahane's inequality 9.2. We begin with the Brunn-Minkowsky inequality (see also Appendix I).

III.1. THEOREM: *Let A, B be two compact sets in $\mathbb{R}^n$. Then*

$$Vol(A+B)^{1/n} \geq Vol(A)^{1/n} + Vol(B)^{1/n} .$$

PROOF: By a simple approximation procedure, we may assume that each of A and B is a union of finitely many disjoint sets, each of which is a product of intervals. The proof is by induction on the total number k of such rectangular boxed in A and B. If $k=2$, i.e. if A and B are rectangular boxes with sides of lengths $(a_i)_{i=1}^n$ and $(b_i)_{i=1}^n$ respectively, then

$$Vol(A+B)^{1/n} = \prod_{i=1}^n (a_i+b_i)^{1/n} , \quad Vol(A) = \prod_{i=1}^n a_i^{1/n} , \quad Vol(B) = \prod_{i=1}^n b_i^{1/n} .$$

By the inequality between the geometrical and arithmetical means,

$$\prod_{i=1}^n \left(\frac{a_i}{a_i+b_i}\right)^{1/n} + \prod_{i=1}^n \left(\frac{b_i}{a_i+b_i}\right)^{1/n} \leq \frac{1}{n}\sum_{i=1}^n \frac{a_i}{a_i+b_i} + \frac{1}{n}\sum_{i=1}^n \frac{b_i}{a_i+b_i} = 1 .$$

Assume now that A and B are composed of a total of $k > 2$ rectangular boxes and that the inequality holds for all sets A', B' such that A', B' are composed of a total of at most $k-1$ boxes. We may and shall assume that the number of rectangular boxes in A is at least 2. Note that parallel shifts of A and B do not change the volume of A, B or $A+B$. We find such a shift of A with the property that one of the coordinate hyperplanes divides A in such a manner that at least one rectangular box in A is in each side of the plane. A is now divided into two sets A', A'' each of which is a disjoint union of finite number of rectangular boxes and the number of boxes in each of A', A'' is strictly smaller that the number in A. Now shift B parallel to the axes in such a manner that the same hyperplane divides B into B', B'' with $\frac{Vol\ B'}{Vol\ B} = \frac{Vol\ A'}{Vol\ A} = \lambda$.

Each of B', B'' has at most the same number of rectangular boxes as B has.

By the induction hypothesis

$$\begin{aligned} Vol(A+B) &\geq Vol(A'+B') + Vol(A''+B'') \\ &\geq [(Vol\ A')^{1/n} + (Vol\ B')^{1/n}]^n + [(Vol\ A'')^{1/n} + (Vol\ B'')^{1/n}]^n \\ &= \lambda[(Vol\ A)^{1/n} + (Vol\ B)^{1/n}]^n + (1-\lambda)[(Vol\ A)^{1/n} + (Vol\ B)^{1/n}]^n \\ &= [(Vol\ A)^{1/n} + (Vol\ B)^{1/n}]^n \end{aligned}$$

□

III.2. It follows from Theorem III.1 that

$$Vol(\lambda A + (1-\lambda)B) \geq (Vol\ A)^{\lambda}(Vol\ B)^{1-\lambda}$$

for all compact sets A, B in $\mathbb{R}^n$ and all $0 < \lambda < 1$. Indeed

$$Vol(\lambda A + (1-\lambda)B)^{1/n} \geq \lambda(Vol\ A)^{1/n} + (1-\lambda)(Vol\ B)^{1/n} \geq \{(Vol\ A)^{\lambda}(Vol\ B)^{1-\lambda}\}^{1/n} .$$

If K is any convex body of finite volume in $\mathbb{R}^n$, put $\mu_K(A) = \frac{Vol(A\cap K)}{Vol\ K}$. Then μ_K clearly satisfies the same kind of inequality

$$\mu_K(\lambda A + (1-\lambda)B) \geq \mu_K(A)^{\lambda}\mu_K(B)^{1-\lambda} .$$

III.3. THEOREM (C. Borell). *Let μ be any Borel probability measure on $\mathbb{R}^n$ satisfying $\mu(\lambda A + (1-\lambda)B) \geq \mu(A)^{\lambda}\mu(B)^{1-\lambda}$. Then for all symmetric convex sets $A \subseteq \mathbb{R}^n$ with $\mu(A) = \theta > \frac{1}{2}$*

$$\mu((tA)^c) \leq \theta\left(\frac{1-\theta}{\theta}\right)^{(1+t)/2} \quad \textit{for all} \quad t > 1 .$$

PROOF. We have the inclusion

$$\mathbb{R}^n \backslash A \supseteq \frac{2}{t+1}(\mathbb{R}^n \backslash tA) + \frac{t-1}{t+1}A$$

(check!). Consequently

$$1-\theta = \mu(A^c) \geq \mu((tA)^c)^{2/(t+1)}\theta^{(t-1)/(t+1)} .$$

□

III.4. Let now $(X, \|\ \ \|)$ be any normed space and let $x_1, \ldots, x_n \in X$. Assume

$$\int_K \|\sum_{i=1}^n a_i x_i\| d\mu(a) = 1$$

where μ is the Lebesgue measure on the cube $K = [-1,1]^n$ normalized as to give $\mu(K) = 1$. Let

$$A = \{a \in K;\ \|\sum_{i=1}^n a_i x_i\| \leq 3\} .$$

Then A is convex symmetric and $\mu(A) \geq \frac{2}{3}$, (since $3\mu(A^c) \leq \int_K \|\Sigma a_i x_i\| d\mu(a) = 1$). Applying Theorem III.3 we get, for all $t > 1$,

III.4.1.
$$\mu\{\|\Sigma_{i=1}^n a_i x_i\| > 3t\} \leq \frac{2}{3}(\frac{1}{2})^{(1+t)/2}$$

and consequently, for all $p > 1$,

$$\int \|\sum_{i=1}^n a_i x_i\|^p d\mu(a) = p \int_0^\infty t^{p-1} \mu(\|\sum_{i=1}^n a_i x_i\| > t) dt$$
$$\leq p\, 3^{p-1} + \frac{2p}{3} \int_3^\infty t^{p-1} (\frac{1}{2})^{(3+t)/6} dt \leq K_p^p$$

for some constant K_p depending only on p. It follows by homogeneity that

III.4.2.
$$(\int_K \|\Sigma_{i=1}^n a_i x_i\|^p d\mu(a))^{1/p} \leq K_p \int_K \|\Sigma_{i=1}^n a_i x_i\| d\mu(a) \ .$$

Kahane's inequality now easily follows from III.4.2. Indeed, by the 1-unconditionality of the Rademacher functions,

$$(\int_K \|\sum_{i=1}^n a_i x_i\|^p d\mu(a))^{1/p} = (\int_0^1 \int_K \|\sum_{i=1}^n a_i r_i(t) x_i\|^p d\mu(a) dt)^{1/p}$$
$$\leq (\int_0^1 \|\sum_{i=1}^n r_i(t) x_i\|^p dt)^{1/p} \ .$$

By the fact that $(a_i)_{i=1}^n$ and $(|a_i| r_i(t))_{i=1}^n$ have the same distribution and by the triangle inequality

$$\int_K \|\sum_{i=1}^n a_i x_i\|^p d\mu(a))^{1/p} = (\int_0^1 \int_K \|\sum_{i=1}^n |a_i| r_i(t) x_i\|^p d\mu(a) dt)^{1/p}$$
$$\geq (\int_0^1 \| \int_K (\sum_{i=1}^n |a_i| r_i(t) x_i) d\mu(a)\|^p dt)^{1/p} = \frac{1}{2}(\int_0^1 \|\sum_{i=1}^n r_i(t) x_i\|^p)^{1/p} \ .$$

We can now conclude from III.4.2 that, for $p > 1$,

$$(\int_0^1 \|\sum_{i=1}^n r_i(t) x_i\|^p dt)^{1/p} \leq 2K_p \int_0^1 \|\sum_{i=1}^n r_i(t) x_i\|$$

which proves Kahane's inequality. (The other side inequality is trivial.)

We remark that this proof does not give the right order of magnitude for the growth rate of the constant. It only gives $K_p \leq Kp$ with $K < \infty$ absolute. See [L.T2] for a proof which gives the right order of magnitude for the constant, $K_p \approx \sqrt{p}$.

APPENDIX IV
PROOF OF THE BEURLING-KATO THEOREM 14.4

Define

$$Ax = \lim_{t \to 0} \frac{S_t x - x}{t}$$

whenever the limit exists. Then A is a linear (usually unbounded) closed operator with dense domain $D(A)$. For any $x \in D(A)$

$$\frac{dS_t x}{dt} = \lim_{s \to 0} \frac{S_{t+s}x - S_t x}{s}$$

exists and

$$S_t x = AS_t x = S_t Ax \ .$$

Let $\sup_t \|I - S_t\| = \rho_0$ and let $\rho_0 < \rho < 2$, say $\rho = \frac{2+\rho_0}{2}$.

STEP 1: For all $\beta \geq \rho$, $((\beta - 1)I + S_t)^{-1}$ exists and $\|((\beta - 1)I + S_t)^{-1}\| \leq \frac{M}{\beta}$ with $M = M(\rho_0)$.

Indeed, since $\|\frac{I-S_t}{\beta}\| \leq \frac{\rho_0}{\rho}$, $T = \Sigma_{n=0}^{\infty}(\frac{I-S_t}{\beta})^n$ converges and its norm is at most $\Sigma_{n=0}^{\infty}(\frac{\rho_0}{\rho})^n = M$. Also,

$$T = (I - \frac{I - S_t}{\beta})^{-1} = \beta((\beta - 1)I + S_t)^{-1} \ .$$

Thus,

$$\|((\beta - 1)I + S_t)^{-1}\| \leq \frac{M}{\beta} \ .$$

STEP 2: There exists $0 < \theta = \theta(\rho_0) \leq \frac{\pi}{2}$ and $M' = M'(\rho_0)$ such that for $z \in \mathbb{C}$ with $|arg\ z| \geq \frac{\pi}{2} - \theta$, $zI + A$ has a bounded inverse and

$$\|(zI + A)^{-1}\| \leq \frac{M'}{|z|} \ .$$

PROOF: Note that for all $t \geq 0$ and $x \in D(A)$,

$$(e^{zt}S_t - I)x = \int_0^t \frac{d}{ds}(e^{zs}S_s)x ds = \int_0^t e^{zs}S_s(z + A)x ds \ .$$

Fix $z = a + ib$. If one can choose t such that $e^{-zt} = 1 - \beta$ with β real and $\beta \geq \rho$, then $(e^{zt}S_t - I)^{-1}$ exists and

$$(zI + A)^{-1} = (e^{zt}S_t - I)^{-1} \int_0^t e^{zs} S_s ds \; .$$

Choose $t = \frac{\pi}{|b|}$. Then,

$$e^{-zt} = e^{-a\pi/|b|} e^{i\pi sign\; b} = -e^{-a\pi/|b|} \; .$$

If $\beta = 1 + e^{-a\pi/|b|}$, then $\beta \geq \rho$ as long as $\frac{|a|}{|b|}$ is small enough, i.e., as long as $|arg\; z| > \frac{\pi}{2} - \theta$ with θ small enough.

For such z and t we get

$$\begin{aligned} \|(zI + A)^{-1}\| &\leq |e^{-zt}| \; \|(S_t - e^{-zt}I)^{-1}\| \int_0^t |e^{zs}| ds \\ &\leq e^{-at} \frac{M}{\beta} \int_0^t e^{as} ds \\ &= e^{-at} \frac{M}{\beta} \frac{e^{at} - 1}{a} = \frac{M}{\beta} \frac{1 - e^{-at}}{a} \\ &\leq \begin{cases} \frac{M'}{\beta} inf(t, \frac{1}{a}) & a \geq 0 \\ M' \; inf(t, \frac{1}{|a|}) & a \leq 0 \, , \end{cases} \end{aligned}$$

so that,

$$\|(zI + A)^{-1}\| \leq M' \; inf(\frac{1}{|a|}, \frac{1}{|b|}) \leq \frac{M''}{|z|} \; .$$

STEP 3: For Γ in the illustration

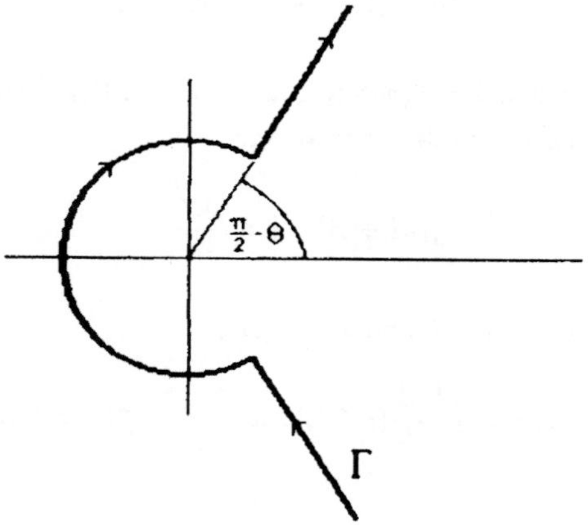

define, for ξ with $|arg\ \xi| < \frac{\theta}{2}$,

$$S_\xi = -\frac{1}{2\pi i}\int_\Gamma e^{-z\xi}(zI + A)^{-1}dz\ .$$

Since $|arg\ z\xi| \leq \frac{\pi}{2} - \theta + \frac{\theta}{2} = \frac{\pi}{2} - \frac{\theta}{2}$ on the infinite rays, $|e^{-z\xi}| \leq e^{-\alpha|z|}$ for some positive α so that the integral converges. One checks that it defines a holomorphic semigroup which extends S_t. To show that it extends S_t compute its derivative,

$$\begin{aligned}\frac{d}{dt}(-\frac{1}{2\pi i}\int_\Gamma e^{-zt}(zI + A)^{-1}dz) &= -\frac{1}{2\pi i}\int_\Gamma e^{-zt}(-z)(zI + A)^{-1}dz \\ &= \frac{1}{2\pi i}\int_\Gamma e^{-zt}(zI + A - A)(zI + A)^{-1}dz \\ &= \frac{1}{2\pi i}\int_\Gamma e^{-zt}(I - A(zI + A)^{-1}dz \\ &= \frac{1}{2\pi i}\int_\Gamma e^{-zt}dt - \frac{A}{2\pi i}\int_\Gamma e^{-zt}(zI + A)^{-1}dz \\ &= A(-\frac{1}{2\pi i}\int_\Gamma e^{-zt}(zI + A)^{-1}dz)\end{aligned}$$

and we get that the two semigroups have the same generator. This of course means that they are equal.

To check that S_ξ is a semigroup, notice that $S_\xi = \frac{-1}{2\pi i}\int_{\Gamma'} e^{-z\xi}(zI + A)^{-1}dz$ for Γ' being, say, $\Gamma - 2$. Then,

$$S_\xi S_{\xi'} = \frac{1}{(2\pi i)^2}\int_\Gamma\int_{\Gamma'} e^{-z\xi - z'\xi'}(zI + A)^{-1}(z'I + A)^{-1}dz'dz\ .$$

By the resolvent formula,

$$S_\xi S_{\xi'} = \frac{1}{(2\pi i)^2}\int_\Gamma\int_{\Gamma'} e^{-z\xi - z'\xi'}(z' - z)^{-1}((zI + A)^{-1} - (z'I + A)^{-1})dz'dz\ ,$$

and by the Cauchy formula,

$$\begin{aligned}S_\xi S_{\xi'} &= \frac{-1}{2\pi i}\int_\Gamma e^{-z\xi - z\xi'}(zI + A)^{-1}dz \\ &= S_{\xi+\xi'}\ .\end{aligned}$$

Finally, we have to check that $\|S_\xi\|$ is bounded in the sector $\{|arg\ \xi| < \frac{\theta}{2}\}$. We leave this to the reader (replace the unit circle with a larger one).

APPENDIX V
THE CONCENTRATION OF MEASURE PHENOMENON FOR GAUSSIAN VARIABLES

We bring here a simple proof, due to Maurey and Pisier, of Theorem V.1 below which in turn is equivalent to Levy's Lemma 2.3 (up to modification of the constants). We also prove a proposition (V.4) relating the deviation from the mean to the deviation from the median.

V.1. THEOREM: *Let $F : \mathbb{R}^n \to \mathbb{R}$ be a Lipschitz function with constant σ ($\mathbb{R}^n$ is endowed with the euclidean metric). Let $g_1, \ldots, g_n$ be independent, mean zero, normalized in L_2, gaussian variables. Then,*

$$P(|F(g_1,\ldots,g_n) - EF(g_1,\ldots,g_n)| > C) \leq 2\ exp(\frac{-2C^2}{\pi^2\sigma^2}) \ .$$

PROOF: We may and shall assume that F is continuously differentiable. Let $H = (h_1, \ldots, h_n)$ be another sequence with the same distribution as $G = (g_1, \ldots, g_n)$ and independent of it. For each $0 \leq \theta \leq \frac{\pi}{2}$ put $G_\theta = G\ sin\ \theta + H\ cos\ \theta$. Then, by the invariance of the gaussian distribution under orthogonal transformations, G_θ and $\frac{d}{d\theta}G_\theta = G\ cos\ \theta - H\ sin\ \theta$ have the same joint distribution as G and H. Consequently, for any convex function φ

$$\begin{aligned} E\varphi(F(G) - EF(\cdot)) \leq E\varphi(F(G) - F(H)) &= E\varphi(\int_0^{\pi/2} \frac{d}{d\theta}F(G_\theta)d\theta) \\ &= E\varphi(\int_0^{\pi/2} (grad\ F(G_\theta), \frac{d}{d\theta}G_\theta)d\theta) \\ &\leq \frac{2}{\pi}\int_0^{\pi/2} E\varphi(\frac{\pi}{2}(grad\ F(G_\theta), \frac{d}{d\theta}G_\theta))d\theta \\ &= E\varphi(\frac{\pi}{2}(grad\ F(G), H)) \ . \end{aligned}$$

For any $\lambda \in \mathbb{R}$ we get

$$E\ exp(\lambda(F(G) - EF(\cdot))) \leq E\ exp(\lambda\frac{\pi}{2}\sum_{i=1}^{n}\frac{\partial F}{\partial x_i}(G)\cdot h_i) \ .$$

Integrating first with respect to the h_i and using the fact that

$$E\ e^{t\Sigma a_i h_i} = E\ e^{t(\Sigma a_i^2)^{1/2}h_1} = e^{t^2\Sigma a_i^2/2}$$

we get

$$E\ exp(\lambda(F(G) - EF(\cdot))) \leq E\ exp\left(\frac{\lambda^2\pi^2}{8}\sum_{i=1}^{n}\frac{\partial F}{\partial x_i}(G)^2\right) .$$

Now, $\Sigma_{i=1}^{n}(\frac{\partial F}{\partial x_i}(x))^2 \leq \sigma^2$ and we get

$$E\ exp(\lambda(F(G) - EF(\cdot))) \leq exp(\lambda^2\pi^2\sigma^2/8) .$$

The rest of the proof is now standard.

□

V.2. Next we show how to deduce from Theorem V.1 a similar theorem for S^{n-1}.

COROLLARY: *Let $f: S^{n-1} \to \mathbb{R}$ be a function with Lipschitz constant σ. Then*

$$\mu(|f - \int f d\mu| > C) \leq 4e^{-\delta C^2 n/\sigma^2}$$

where μ is the Haar measure on S^{n-1} and $\delta > 0$ an absolute constant.

PROOF: First notice that, by the invariance of the canonical gaussian measure on $\mathbb{R}^n$ under orthogonal transformations,

$$\frac{1}{(\Sigma_{i=1}^{n} g_i^2)^{1/2}}(g_1, g_2, \ldots, g_n)$$

has the same distribution as $a = (a_1, \ldots, a_n)$ does with respect to $d\mu(a)$.

Given a function $f: S^{n-1} \to \mathbb{R}$ with Lipschitz constant σ and $\int f d\mu = 0$, define $\tilde{f}: \mathbb{R}^n \to \mathbb{R}$ by $\tilde{f}(x) = \|x\| f(\frac{x}{\|x\|})$, where $\|x\| = (\Sigma\ x_i^2)^{\frac{1}{2}}$. Then it is easily checked that the Lipschitz constant of $\tilde{f}$ is at most 3σ.

Now, for any $\delta > 0$,

$$\begin{aligned}
\mu(|f| > C) &= P(|f(\frac{g}{\|g\|})| > C) \\
&= P(\|g\|^{-1}|\tilde{f}(g)| > C) = P(|\tilde{f}(g)| > C\|g\|) \\
&\leq P(|\tilde{f}(g)| > \delta C\sqrt{n}) + P(\|g\| < \delta\sqrt{n}) \\
&\leq P(|\tilde{f}(g)| > \delta C\sqrt{n}) + P(|\|g\| - E\|g\|| > E\|g\| - \delta\sqrt{n}).
\end{aligned}$$

A simple computation shows that $E\|g\| > 2\delta\sqrt{n}$ for some $\delta > 0$ absolute. For that δ we get from Theorem V.1, applied to $\tilde{f}$ and to $\|\cdot\|$, that

$$\mu(|f| > C) \leq 2e^{-\frac{2\delta^2 C^2 n}{\pi^2 9\sigma^2}} + 2e^{-\frac{2\delta^2 n}{\pi^2}} .$$

Since we may assume $C \leq \sigma$, we get that the first term is dominating and the result follows.

□

V.3. REMARK: The last corollary evaluates the probability of a large deviation of f from its expectation while the results in Section 2 deals with deviation from the median. This is another advantage of the method here. It saves us the trouble of passing from the median to the average (Lemma 5.1). Also, as we shall see below, it is not hard to get, using Theorem V.1, a similar estimate for the deviation from the median.

V.4. We conclude this appendix with a proposition showing that estimation of large deviation from the median is equivalent to the estimation of large deviation from the expectation.

PROPOSITION: *Each of the following four conditions implies the other three, where the constants (K_i, δ_i) depend linearly each on the other.*

1. $\exists A$ *s.t.* $\forall C$

$$P(|f - A| > C) \leq K_1 e^{-\delta_1 C^2}$$

2. $\forall C$

$$P(|f - \tilde{f}| > C) \leq K_2 e^{-\delta_2 C^2}$$

where f and $\tilde{f}$ are independent and identically distributed.

3. $\forall C$

$$P(|f - Ef| > C) \leq K_3 e^{-\delta_3 C^2}$$

4. $\forall C$

$$P(|f - Mf| > C) \leq K_4 e^{-\delta_4 C^2}$$

(Mf is the median of f).

More precisely,

$$K_1 \leq K_4 \leq 2K_3 \leq 2K_2 \leq 4K_1$$

and

$$\delta_1 \geq \delta_4 \geq \frac{\delta_3}{4} \geq \frac{\delta_2}{8} \geq \frac{\delta_1}{32} .$$

PROOF: $1 \Longrightarrow 2$

$$P(|f - \tilde{f}| > C) \leq P(|f - A| > \frac{C}{2}) + P(|\tilde{f} - A| > \frac{C}{2})$$
$$\leq 2K_1 e^{-\delta_1 C^2/4}$$

$2 \Longrightarrow 3$

$$E\, exp(\lambda^2 |f - \tilde{f}|^2) = \int_0^\infty 2\lambda^2 C e^{\lambda^2 C^2} P(|f - \tilde{f}| > C) dC$$
$$\leq 2K_2 \lambda^2 \int_0^\infty C e^{\lambda^2 C^2 - \delta_2 C^2} dC .$$

For $\lambda = \sqrt{\frac{\delta_2}{2}}$, we get

$$E\ exp\left(\frac{\delta_2|f-\tilde{f}|^2}{2}\right) \leq K_2\ .$$

Now, $\varphi(t) = e^{\delta_2 t^2/2}$ is convex, so (as in the proof of V.1),

$$E\ exp(\frac{\delta_2}{2}|f-Ef|^2) \leq K_2$$

and

$$\begin{aligned} P(|f-Ef|>C) &\leq E\ exp(\frac{\delta_2}{2}|f-Ef|^2 - \frac{\delta_2}{2}C^2) \\ &\leq K_2 e^{-\frac{\delta_2}{2}C^2}\ . \end{aligned}$$

$3 \Longrightarrow 4$ is similar to the end of the proof 7.5: Let $C_0 = \sqrt{\frac{\log 2K_3}{\delta_3}}$. Then

$$P(|f-Ef|>C_0) \leq \frac{1}{2}\ .$$

In particular,

$$P(f > C_0 + Ef),\ P(f < Ef - C_0) \leq \frac{1}{2}$$

so that

$$Ef - C_0 < Mf < C_0 + Ef\ .$$

Now, for $C \geq 2C_0$,

$$\begin{aligned} P(|f-Mf|>C) &\leq P(|f-Ef|>C-C_0) \\ &\leq K_3 e^{-\delta_3 (C-C_0)^2}. \\ &\leq K_3 e^{-\frac{\delta_3}{4}C^2}\ . \end{aligned}$$

For $C < 2C_0$,

$$e^{-\frac{\delta_3}{4}C^2} \geq e^{-\delta_3 C_0^2} = (2K_3)^{-1}$$

and

$$P(|f-Mf|>C) \leq 1 \leq 2K_3 e^{-\frac{\delta_3 C^2}{4}}\ ,$$

so that 4 holds with $\delta_4 = \frac{\delta_3}{4}$ and $K_4 = 2K_3$.

$4 \Longrightarrow 1$ is trivial.

$\square$

NOTES AND REMARKS

2.1 The theorem is stated in Levy's [Lev] with a proof which has not been understood for a long time. A complete proof appears in [Schm]. In Appendix I a proof (of Gromov) is given which is based on Levy's original idea. See also [F.L.M.] for a short proof.

2.3 is from [Lev].

2.4–2.8 are from [M1].

3.1 There are a substantial number of results about distances between different finite dimensional normed spaces. A detailed study of these problems can be found in [T-J1]. We only mention a few results. Let $\mathcal{K}_n$ be the family of all n-dimensional normed spaces equipped with the Banach-Mazur distance. Theorem 3.3 implies that the diameter of $\mathcal{K}_n$ is $\leq n$. It was shown by E. Gluskin [Gl] that *diam* $\mathcal{K}_n \geq cn$ for some absolute constant $c > 0$. However, when restricted to some natural subclasses, the diameter is much smaller. For example, N. Tomczak-Jaegermann [T-J2] has proved that the distance between any two spaces with 1-symmetric bases in $\mathcal{K}_n$ is at most $C\sqrt{n}$ for some absolute $C < \infty$. Recently J. Lindenstrauss and A. Szankowski [L.S.] have shown that for some (specific) $\alpha < 1$ and $C < \infty$ the distance between any two members of $\mathcal{K}_n$ with 1-unconditional bases is at most Cn^{α}. (The α computed in [L.S.] is $\geq \frac{5}{9}$, while one naturally expects the right α to be $\frac{1}{2}$ or $\frac{1}{2} + \varepsilon$ with $C = C(\varepsilon)$.) A similar estimate was obtained in [B.M.1] for the distance between $X \in \mathcal{K}_n$ and its dual, X^*. In another direction, subfamilies of $\mathcal{K}_n$ with some conditions on their type and cotype were considered in [Kw1], [F.L.M.], [T-J.3], [D.M.T-J.] and [B.G.] where again estimates of order better than n were given to their diameters. Theorem 3.3 also implies that the radius of $\mathcal{K}_n$ is equal to $\sqrt{n}$ with ℓ_2^n in (one of?) the center(s). If $X \in \mathcal{K}_n$ is of almost extreme distance from ℓ_2^n, i.e., $d(\ell_2^n, X) \geq cn$ ($c > 0$ fixed), then it was proved in [M.W.] that X necessarily contains a $(1+\varepsilon)$-isomorphic copy of ℓ_1^k with $k = k(c, \varepsilon, n) \to \infty$ as $n \to \infty$.

3.3 is taken from F. John's [Jo].

3.4 This is a weak form of the Dvoretzky-Rogers Lemma [D.R.] (or see [Da]). The estimate can be improved to $1 \geq \|x_i\| \geq \frac{(n-i+1)^{1/2}}{n^{1/2}+(i-1)^{1/2}}$. The proof given here was shown to us by W. Johnson.

4.2, 4.3 were proved in [M1]. The estimates on the constant $c(\varepsilon)$ in 4.2 Remark a, and 4.3, can be improved to $c(\varepsilon) \geq c\varepsilon^2$. This follows from a recent proof of Dvoretzky's Theorem due to Y. Gordon [Go].

4.4–4.7 are proved in [F.L.M.].

4.8–4.11 are taken from [M3]. The best constant in 4.8 is $\frac{c\sqrt{\phi}}{M_{r^*}}|x| \leq \|x\|$ (see P.T-J.]).

5.1–5.4 are taken from [F.L.M.] except for Fact 2 of **5.4** which was proved in [B.D.G.J.N.]. The condition $a \cdot b \leq \sqrt{n}$ in **5.1** is not needed (T. Figiel, private communication). Actually all of Lemma 5.1 is not needed, instead one can use Proposition V.4.

5.3 admits a lot of generalizations, we indicate here only one direction: Kashin [Kas] has proved that for any $\lambda < 1$, ℓ_1^n contains a $C(\lambda)$-isomorphic copy of ℓ_2^k, for $k = [\lambda n]$. The proof uses a different approach involving the following property of ℓ_1^n: Let O be the ellipsoid of maximal volume inscribed in the unit ball, K, of ℓ_1^n, then $vr(K) = (Vol\ K/Vol\ O)^{1/n}$ is bounded independently of n (see also [Sz2]). The quantity $vr(K)$, called the volume ratio of K, was introduced and studied for various K's in [S.T-J.], where a generalization of Kashin's theorem is proved for spaces whose unit balls have bounded volume ratio. For recent developments on this subject (in particular its relation with cotype 2), see [B.M.2] and [Mi.P.].

5.5 For the original proof of Khinchine's inequality see [L.T.1]. The best constants A_p, B_p were computed, for $p = 1$, by S. Szarek [Sz1] ($A_1 = \frac{1}{\sqrt{2}}$), and for all the p's by U. Haagerup [Haa].

5.7 was first proved in [M1]. There is a disturbing gap between $k(\ell_\infty^n) \approx log\ n$ and $k(\ell_q^n) \approx n^{2/q}$ (which becomes a constant for $q = log\ n$ for which $d(\ell_q^n, \ell_\infty^n) = e$). This gap can be bridged – one can prove $k(\ell_q^n) \geq cqn^{2/q}$ for all $q \leq log\ n$ with an absolute $c > 0$. This estimate is best possible up to numerical constants.

5.8 The original proof of Dvoretzky's Theorem only gives $k(X) \geq c\sqrt{log\ n/log\ log\ n}$. As stated here, the theorem was first proved in [M1] (with the proof given here).

5.9 The theorem (with a different dependence on λ) was proved in [M2]. We follow [M3].

6.1 The notion of Levy family was introduced in [G.M.] which also contains a list of examples of Levy families and applications of the concentration phenomenon of such families to topology and fixed point theorems.

6.2 was observed and used in [A.M.].

6.3 was proved and used for the Local Theory by Maurey [Ma].

6.5–6.7 These examples were observed in [G.M.]. Concentration properties for $W_{n,k}$ and $G_{n,k}$ were also noted and applied in [M4], [M5].

6.9 was proved in [G.M.]. This method was later extended in [Al.M.] to discrete cases.

7.4 is due to K. Azuma (see [St] for this and related inequalities). For other related inequalities, see Lemma 8.4 below and Proposition 3.1 in [J.S.Z.].

7.5 was proved by B. Maurey [Ma]. The proof here follows [Sc1]. We remark that unlike the similar case of E_2^n no precise isoperimetric inequality for Π_n is known.

7.7–7.8 is taken from [Sc1].

7.9 For a proof through an isoperimetric inequality, see L.H. Harper's [Har] (and [A.M.1] for the form in which it appears here). [Fr.F] contains a simpler proof of Harper's isoperimetric inequality.

7.15 is taken from [Sc2], which contains also generalizations to spaces other than ℓ_1^n.

8.1 p-stables were introduced by P. Levy. Their relevance to the problem of embedding ℓ_p in L_r was noticed by M.I. Kadec.

8.2–8.8 is taken from [J.S.] in which a more general case is also considered (in the statement of Theorem 8.8 one can replace ℓ_1^n by ℓ_r^n for any $0 < r \leq 1$ and also by $1 < r < p$ provided $1 + \varepsilon$ is replaced with $K_{r,p}$ and $\beta = \beta(r,p)$). Theorem 8.8 was further generalized, replacing ℓ_1^n by a wide family of spaces, by G. Pisier [Pis2], see Chapter 13. For generalizations in a different direction, replacing ℓ_p^m by a general finite dimensional subspace of L_p see [Sc3], [Sc4].

8.4 This inequality is due to G. Pisier.

9.1 The type and cotype inequalities were first considered by J. Hoffmann-Jorgensen [H-J] in connection with limit theorems for independent Banach space valued random variables. [M.P.] contains the first major relations between these inequalities and geometrical properties of Banach spaces.

9.2 is proved in [Kah]. The original proof as well as the proof in Appendix III do not give the right order of the constant K_p. For a proof, due to Kwapien, which gives the right order $K_p \sim \sqrt{p}$ see [Kw2] or [L.T.2].

9.6 is taken from [F.L.M.].

9.7 appears here for the first time. It should be compared with Theorem 13.12 below.

9.10 The Rademacher projection, Rad_n, and its relevance to duality between type and cotype was noticed in [M.P.].

10. is based on [A.M.1] and [A.M.2]. However, the main idea of the proof of Theorem 10.7 is an adaptation of Krivine's original proof [Kr] to the finite dimensional case.

11. A good reference for combinatorial treatment of Ramsey's Theorem is [G.R.S.]. For a review of applications of Ramsey's Theorem to Banach space theory see Odell's [Od]. A.

Brunel and L. Sucheston were the first to notice the relevance of Ramsey's Theorem to Banach space theory; they proved Theorem **11.4** (cf.[B.S.1], [B.S.2]).

11.8 The Lemma here is due to James [Ja].

12.3–12.4 are due to J.L. Krivine [Kr]. We follow here a simpler proof due to H. Lemberg [Lem].

The proof given for **12.5** appears here for the first time. It simplifies the proof of the Maurey-Pisier Theorem 13.2.

13.2 is due to Maurey and Pisier [M.P.]. We follow [M.S.] (for the cotype case, 13.8-13.9) and [Pis2] (for the type case, 13.10-13.12). [Sc3] and [Sc4] contain some extensions of **13.12**: It is proved there that any k-dimensional subspace of L_p, $1 < p < 2$, $(1+\varepsilon)$ imbeds into X provided $k^{1+1/p}(\log k)^{-1} \leq \delta(p,\varepsilon)(ST_p(X))^q$. It is not known if the conclusion holds with $k \leq \delta(p,\varepsilon)(ST_p(X))^q$.

14.5 is contained in [Pis1].

14.6 is contained in [Pis4] and [Pis5] with a different proof. The proof here was shown to us by T. Figiel. The estimate $\|Rad_n X\| \leq K \log n$ for X with $\dim X = n$ is best possible, up to the choice of K. This was shown by J. Bourgain [Bou]. But in special cases (notably, spaces with 1-unconditional bases) one gets $\|Rad_n X\| \leq K\sqrt{\log n}$, see [Pis5].

15. The material in this chapter is due mainly to T. Figiel and N. Tomczak-Jaegermann [F.T-J.].

15.2 is a generalization, due to D. Lewis [Lew], of a theorem of F. John [Jo].

I.1.5.B first appear in that form in [Gr1]. It is more commonly referred to as the Levy-Gromov Theorem.

II.1 The theorem is due to Maurey and Pisier [M.P.]. The proof here is different.

II.2–II.3 For more information about p-convexity, q-concavity, upper p-estimate and lower q-estimate, see [L.T.2].

III.3 is from [Bo].

III.4 G. Pisier showed us how to deduce Kahane's inequality from Borell's theorem.

IV. The proof is taken from [R.S.].

V.1 See [Pis7] for this and some generalizations.

V.4 is probably known to several people but we don't know of any reference.

INDEX

(Roman numerals refer to appendices)

REFERENCES

[Al] F. Almgren, Existence and regularity almost everywhere of solutions to elliptic variational problems with constraints, Mem. A.M.S., 4 (1976).

[Al.M.] N. Alon and V.D. Milman, λ_1, isoperimetric inequalities for graphs, and superconcentrators, J. Comb. Theory, Ser. B, 38 (1985), 73-88.

[A.M.1] D. Amir and V.D. Milman, Unconditional and symmetric sets in n-dimensional normed spaces, Israel J. Math., 37 (1980), 3-20.

[A.M.2] D. Amir and V.D. Milman, A quantitative finite-dimensional Krivine theorem, Israel J. Math., 50 (1985), 1-12.

[B.D.G.J.N.] G. Bennett, L.E. Dor, V. Goodman, W.B. Johnson and C.M. Newman, On uncomplemented subspaces of L_p,$1 < p < 2$, Israel J. Math., 26 (1977), 178-187.

[B.G.] Y. Benyamini and Y. Gordon, Random factorization of operators between Banach spaces, J. d'Analyse Math., 39 (1981), 45-74.

[B.G.M.] M. Berger, P. Gauduchon and E. Mazet, Le Spectre d'une Variété Riemannienne, Lecture Notes in Mathematics, Vol. 194, Springer 1971.

[Ber] P.M. Bérard, Lectures on spectral geometry, IMPA 1985, Rio de Janeiro. Lecture Notes in Mathematics, Vol. 1207, Springer 1986.

[Be] A. Beurling, On analytic extension of semigroups of operators, J. Funct. Anal., 6 (1970), 387-400.

[Bo] C. Borell, The Brunn-Minkowski inequality in Gauss spaces, Inventiones Math., 30 (1975), 207-216.

[Bou] J. Bourgain, On martingales transforms in finite dimensional lattices with an appendix on the K-convexity constant, Math. Nachrichten, 119 (1984), 41-53.

[B.M.1] J. Bourgain and V.D. Milman, Distances between normed spaces, their subspaces and quotient spaces, Integral Equations & Operator Theory, 9 (1986), 31-46.

[B.M.2] J. Bourgain and V.D. Milman, On Mahler's conjecture on the volume of a convex symmetric body and its polar, Preprint, IHES 1985.

[B.S.1] A. Brunel and L. Sucheston, On B-convex Banach spaces, Math. Systems Th., 7 (1974), 294-299.

[B.S.2] A. Brunel and L. Sucheston, On J-convexity and some ergodic super-properties of Banach spaces, Trans. A.M.S., 204 (1975), 79-90.

[Br] H. Brunn, Über Ovale und Eiflächen, Inaug. Diss., München, 1887.

[Bu.Ma.] Yu. Burago, and V. Mazya, Some problems of the potential theory and the function theory for domains with non-regular boundary, Zapiski LOMI, Leningrad, 1967 (Russian).

[Bus] P. Buser, On Cheeger's inequality $\lambda_1 \geq \frac{h^2}{4}$, Preprint.

[Ch] J. Cheeger, A lower bound for the smallest eigenvalue of the Laplacian, Problems in Analysis, Symp. in honor of Bochner, Princeton, 1970, 195-199.

[C.E.] J. Cheeger and E.G. Ebin, Comparison Theorems in Riemannian Geometry, North-Holland, Amsterdam, 1975.

[D.M.T-J.] W.J. Davis, V.D. Milman and N. Tomczak-Jaegermann, The distance between certain n-dimensional spaces, Israel J. Math., 39 (1981), 1-15.

[Da] M.M. Day, Normed Linear Spaces, Springer, 1973.

[Do] W.F. Donoghue Jr., Distributions and Fourier Transforms, Academic Press, 1969.

[D.S.] N. Dunford and J.T. Schwartz, Linear Operators, Part I, Wiley, 1957.

[Dv.] A. Dvoretzky, Some results on convex bodies and Banach spaces, Proc. Symp. on Linear Spaces, Jerusalem 1961, 123-160.

[D.R.] A. Dvoretzky and C.A. Rogers, Absolute and unconditional convergence in normed linear spaces, Proc. Nat. Acad. Sci., U.S.A., 36 (1950), 192-197.

[F.F.] H. Federer and W. Fleming, Normal and integral currents, Ann. of Math., 72 (1960), 458-520.

[Fe] W. Feller, An Introduction to Probability Theory and its Applications, Vol. II, Wiley, 1966.

[Fi] T. Figiel, A short proof of Dvoretzky's theorem on almost spherical sections, Compositio Math., 33 (1976), 297-301.

[F.L.M.] T. Figiel, J. Lindenstrauss and V.D. Milman, The dimension of almost spherical sections of convex bodies, Acta Math., 139 (1977), 53-94.

[F.T-J.] T. Figiel and N. Tomczak-Jaegermann, Projections onto Hilbertian subspaces of Banach spaces, Israel J. Math., 33 (1979), 155-171.

[Fr.F.] P. Frankl and Z. Füredi, A short proof for a theorem of Harper about Hamming-spheres, Discrete Math., 34 (1981), 311-313.

[Gi] D.P. Giesy, On a convexity condition in normed linear spaces, Trans. A.M.S., 125 (1966), 114-146.

[Gl] E.D. Gluskin, The diameter of the Minkowski compactum is roughly equal to n, Funct. Anal. Appl., 15 (1981), 72-73.

[Go] Y. Gordon, Some inequalities for gaussian processes and applications, Israel J. Math., 50 (1985), 265-289.

[G.R.S.] R.L. Graham, B.L. Rothschild and J.H. Spencer, Ramsey Theory, Wiley, New York, 1980.

[G.K.M.] D. Gromoll, W. Klingenberg and W. Meyer, Riemannsche Geometrie in Gruppen, Springer, 1968.

[Gr1] M. Gromov, Paul Levy's isoperimetric inequality, Preprint, 1980.

[Gr2] M. Gromov, Filling Riemannian manifolds, J. Diff. Geom., 18 (1983), 1-147.

[Gr3] M. Gromov, Partial Differential Relations, Springer Verlag, to appear.

[G.M.] M. Gromov and V.D. Milman, A topological application of the isoperimetric inequality, Amer. J. Math., 105 (1983), 843-854.

[Haa] Uffe Haagerup, The best constants in the Khinchine inequality, Studia Math., 70 (1982), 231-283.

[Had] H. Hadwiger, Vorlesungen über Inhalt, Oberfläche und Isoperimetrie, Springer Verlag, 1957.

[Har] L.H. Harper, Optimal numberings and isoperimetric problems on graphs, J. Comb. Theory, 1 (1966), 385-393.

[H-J] J. Hoffmann-Jorgensen, Sums of independent Banach space valued random variables, Aarhus Universitat, 1972/73.

[Ja] R.C. James, Uniformly non-square Banach spaces, Ann. of Math., 80 (1964), 542-550.

[Jo] F. John, Extremum problems with inequalities as subsidiary conditions, Courant Anniversary Volume, Interscience, New York, 1948, 187-204.

[J.S.] W.B. Johnson and C. Schechtman, Embedding ℓ_p^m into ℓ_1^n, Acta Math., 149 (1982), 71-85.

[J.S.Z.] W.B. Johnson, G. Schechtman and J. Zinn, Best constants in moment inequalities for linear combinations of independent and exchangeable random variables, Ann. Prob., 13 (1985), 234-253.

[Kah] J.P. Kahane, Series of Random Functions, Heath Math. Monographs, Lexington, Mass., Heath & Co., 1968.

[Kas] B.S. Kashin, Sections of some finite dimensional sets and classes of smooth functions, Izv. ANSSSR, ser. mat. 41 (1977), 334-351 (Russian).

[Ka] T. Kato, A characterization of holomorphic semi-groups, Proc. A.M.S., 25 (1970), 495-498.

[Kn] H. Knothe, Contributions to the theory of convex bodies, Michigan Math. J., 4 (1957), 39-52.

[Kr] J.L. Krivine, Sous-espaces de dimension finie des espaces de Banach réticulés, Ann. of Math., 104 (1976), 1-29.

[Kw1] S. Kwapien, Isomorphic characterizations of inner product spaces by orthogonal series with vector coefficients, Studia Math., 44 (1972), 583-595.

[Kw2] S. Kwapien, A theorem on the Rademacher series with vector coefficients, Proc. Int. Conf. on Prob. in Banach Spaces, Lecture Notes in Math. Vol. 526, Springer, 1976.

[Law] B. Lawson, Minimal varieties, Proc. of Symp. in Pure Math., A.M.S., 27, Part I, (1975), 143-177.

[Lem] H. Lemberg, Nouvelle démonstration d'un théorème de J.L. Krivine sur la finie representation de ℓ_p dans un espace de Banach, Israel J. Math., 39 (1981), 341-348.

[L.W.Z.] R. Lepage, M. Woodroofe and J. Zinn, Convergence to a stable distribution via order statistics, Ann. Prob., 9 (1981), 624-632.

[Le] P. Levy, Problèmes concrets d'analyse fonctionnelle, Gauthier Villars, Paris 1951.

[Lew] D.R. Lewis, Ellipsoids defined by Banach ideal norms, Mathematika 26 (1979), 18-29.

[L.S.] J. Lindenstrauss and A. Szankowski, On the Banach-Mazur distance between spaces having an unconditional basis, Aspects of Positivity in Functional Analysis, North Holland, to appear.

[L.T.] J. Lindenstrauss and L. Tzafriri, Classical Banach spaces, Lecture Notes in Mathematics Vol. 338, Springer-Verlag, Berlin, 1973.

[L.T.1.] J. Lindenstrauss and L. Tzafriri, Classical Banach Spaces I, Sequence Spaces, Springer-Verlag, Berlin, 1977.

[L.T.2.] J. Lindenstrauss and L. Tzafriri, Classical Banach Spaces II, Function Spaces, Springer-Verlag, Berlin, 1979.

[Lo] M. Loewe, Probability Theory I, Springer-Verlag, 1977.

[Mar.P.] M.B. Marcus and G. Pisier, Characterizations of almost surely continuous p-stable random Fourier series and strongly stationary processes, Acta Math., 152 (1984), 245-301.

[Ma] B. Maurey, Construction de suites symétriques, C.R.A.S., Paris, 288 (1979), 679-681.

[M.P.] B. Maurey and G. Pisier, Séries de variables aléatoires vectorielles indépendentes et propriétés géométriques des espaces de Banach, Studia Math., 58 (1976), 45-90.

[Maz] V. Mazya, Classes of domains and imbedding theorems for function spaces. Dokl. Ak. Sci. USSR 133 #3, 527-530 (1960), (English Transl. p. 882-884).

[M1] V.D. Milman, A new proof of the theorem of A. Dvoretzky on sections of convex bodies, Func. Anal. Appl., 5 (1971), 28-37 (translated from Russian).

[M2] V.D. Milman, Almost Euclidean quotient spaces of subspaces of finite dimensional normed spaces, Proc. A.M.S., 94 (1985), 445-449.

[M3] V.D. Milman, Random subspaces of proportional dimension of finite dimensional normed spaces: Approach through the isoperimetric inequality, Banach Spaces, Proc. Missouri Conference 1984, Lecture Notes in Mathematics, Vol. 1166, Springer 1985.

[M4] V.D. Milman, Asymptotic properties of functions of several variables defined on homogeneous spaces, Soviet Math. Dokl., 12 (1971), 1277-1281.

[M5] V.D. Milman, On a property of functions defined on infinite-dimensional manifolds, Soviet Math. Dokl., 12 (1971), 1487-1491.

[Mi.P.] V.D. Milman and G. Pisier, Banach spaces with a weak cotype 2 property, Israel J., to appear.

[M.S.] V.D. Milman and M. Sharir, A new proof of the Maurey-Pisier theorem, Israel J. Math., 33 (1979), 73-87.

[M.W.] V.D. Milman and H. Wolfwon, Minkowski spaces with extremal distance from Euclidean space, Israel J. Math., 29 (1978), 113-130.

[Mi] J. Milnor, Morse Theory, Annals of Math. Studies No. 51, 1963, Princeton University Press.

[Od] E. Odell, Applications of Ramsey theorems to Banach space theory, Notes In Banach Spaces, University of Texas Press, Austin and London, 1980.

[P.T-J.] A. Pajor and N. Tomczak-Jaegermann, Subspaces of small codimension of finite-dimensional Banach spaces, preprint.

[Pe] A. Pelczynski, Geometry of finite dimensional Banach spaces and operators ideals, Notes in Banach Spaces, University of Texas Press, Austin and London, 1980.

[Pie] A. Pietsch, Operators Ideals, North Holland, 1978.

[Pis1] G. Pisier, Holomorphic semi-groups and the geometry of Banach spaces, Ann. Math., 115 (1982), 375-392.

[Pis2] G. Pisier, On the dimension of the ℓ_p^n-subspaces of Banach spaces, for $1 \leq p < 2$, Trans. A.M.S., 276 (1983), 201-211.

[Pis3] G. Pisier, Counterexamples to a conjecture of Grothendieck, Acta Math., 151 (1983), 181-208.

[Pis4] G. Pisier, Un théorème sur les opérateurs linéaires entre espaces de Banach qui se factorisent par un espace de Hilbert, Ann.Scient.Ec.Norm. Sup., 13 (1980), 23-43.

[Pis5] G. Pisier, Remarques sur un résultat non publié de B. Maurey, Sém. d'Anal. Fonctionnelle 1980/81, Ecole Polytechnique, Paris.

[Pis6] G. Pisier, Factorization of Linear Operators and the Geometry of Banach Spaces, CBMS Regional Conf. Series in Math., 1986.

[Pis7] G. Pisier, Probabilistic methods in the geometry of Banach spaces, preprint.

[Ra] F.P. Ramsey, On a problem of formal logic, Proc. Lond. Math. Soc. (2), 30 (1929), 264-286.

[R.S.] M. Reed and B. Simon, Methods of Modern Mathematical Physics II, Fourier Analysis, Self Adjointness, Academic Press, 1975.

[Sc1] G. Schechtman, Levy type inequality for a class of metric spaces, Martingale Theory in Harmonic Analysis and Banach Spaces, Springer-Verlag 1981, 211-215.

[Sc2] G. Schechtman, Random embedding of euclidean spaces in sequence spaces, Israel J. Math., 40 (1981), 187-192.

[Sc3] G. Schechtman, Fine embeddings of finite dimensional subspaces of L_p, $1 \leq p < 2$, into ℓ_1^m, Proc. A.M.S., 94 (1985), 617-623.

[Sc4] G. Schechtman, Fine embeddings of finite dimensional subspaces of L_p, $1 \leq p < 2$, into finite dimensional normed spaces II, Longhorn Notes, Univ. of Texas, 1984/85.

[Schm] E. Schmidt, Die Brunn-Minkowski Ungleichung, Math. Nachr., 1 (1948), 81-157.

[St] W.F. Stout, Almost Sure Convergence, Academic Press, 1974.

[Sz1] S.J. Szarek, On the best constant in the Khinchine inequality, Studia Math. 58 (1976), 197-208.

[Sz2] S.J. Szarek, On Kashin almost Euclidean orthogonal decomposition of ℓ_1^n, Bull. Acad. Polon. Sci. 26 (1978), 691-694.

[Szan] A. Szankowski, On Dvoretzky's theorem on almost spherical sections of convex bodies, Israel J. Math., 17 (1974), 325-338.

[T-J.1] N. Tomczak-Jaegermann, Banach-Mazur Distances and Finite Dimensional Operator Ideals, Pitman, To appear.

[T-J.2] N. Tomczak-Jaegermann, The Banach-Mazur distance between symmetric spaces, Israel J. Math., 46 (1983), 40-66.

[T-J.3] N. Tomczak-Jaegermann, The weak distance between finite-dimensional Banach spaces, Math. Nach., 119 (1984), 291-307.

[Vi] N. Ja. Vilenkin, Special functions and the Theory of Group Representations, Translations of Math. Monographs, A.M.S., Vol. 22 (1968).

[Zy] A. Zygmund, Trigonometric Series, Cambridge Univ. Press, 1977.

Vol. 1664: M. Väth, Ideal Spaces. V, 146 pages. 1997.

Vol. 1665: E. Giné, G. R. Grimmett, L. Saloff-Coste, Lectures on Probability Theory and Statistics 1996. Editor: P. Bernard. X, 424 pages, 1997.

Vol. 1666: M. van der Put, M. F. Singer, Galois Theory of Difference Equations. VII, 179 pages. 1997.

Vol. 1667: J. M. F. Castillo, M. González, Three-space Problems in Banach Space Theory. XII, 267 pages. 1997.

Vol. 1668: D. B. Dix, Large-Time Behavior of Solutions of Linear Dispersive Equations. XIV, 203 pages. 1997.

Vol. 1669: U. Kaiser, Link Theory in Manifolds. XIV, 167 pages. 1997.

Vol. 1670: J. W. Neuberger, Sobolev Gradients and Differential Equations. VIII, 150 pages. 1997.

Vol. 1671: S. Bouc, Green Functors and *G*-sets. VII, 342 pages. 1997.

Vol. 1672: S. Mandal, Projective Modules and Complete Intersections. VIII, 114 pages. 1997.

Vol. 1673: F. D. Grosshans, Algebraic Homogeneous Spaces and Invariant Theory. VI, 148 pages. 1997.

Vol. 1674: G. Klaas, C. R. Leedham-Green, W. Plesken, Linear Pro-*p*-Groups of Finite Width. VIII, 115 pages. 1997.

Vol. 1675: J. E. Yukich, Probability Theory of Classical Euclidean Optimization Problems. X, 152 pages. 1998.

Vol. 1676: P. Cembranos, J. Mendoza, Banach Spaces of Vector-Valued Functions. VIII, 118 pages. 1997.

Vol. 1677: N. Proskurin, Cubic Metaplectic Forms and Theta Functions. VIII, 196 pages. 1998.

Vol. 1678: O. Krupková, The Geometry of Ordinary Variational Equations. X, 251 pages. 1997.

Vol. 1679: K.-G. Grosse-Erdmann, The Blocking Technique. Weighted Mean Operators and Hardy's Inequality. IX, 114 pages. 1998.

Vol. 1680: K.-Z. Li, F. Oort, Moduli of Supersingular Abelian Varieties. V, 116 pages. 1998.

Vol. 1681: G. J. Wirsching, The Dynamical System Generated by the 3n+1 Function. VII, 158 pages. 1998.

Vol. 1682: H.-D. Alber, Materials with Memory. X, 166 pages. 1998.

Vol. 1683: A. Pomp, The Boundary-Domain Integral Method for Elliptic Systems. XVI, 163 pages. 1998.

Vol. 1684: C. A. Berenstein, P. F. Ebenfelt, S. G. Gindikin, S. Helgason, A. E. Tumanov, Integral Geometry, Radon Transforms and Complex Analysis. Firenze, 1996. Editors: E. Casadio Tarabusi, M. A. Picardello, G. Zampieri. VII, 160 pages. 1998

Vol. 1685: S. König, A. Zimmermann, Derived Equivalences for Group Rings. X, 146 pages. 1998.

Vol. 1686: J. Azéma, M. Émery, M. Ledoux, M. Yor (Eds.), Séminaire de Probabilités XXXII. VI, 440 pages. 1998.

Vol. 1687: F. Bornemann, Homogenization in Time of Singularly Perturbed Mechanical Systems. XII, 156 pages. 1998.

Vol. 1688: S. Assing, W. Schmidt, Continuous Strong Markov Processes in Dimension One. XII, 137 page. 1998.

Vol. 1689: W. Fulton, P. Pragacz, Schubert Varieties and Degeneracy Loci. XI, 148 pages. 1998.

Vol. 1690: M. T. Barlow, D. Nualart, Lectures on Probability Theory and Statistics. Editor: P. Bernard. VIII, 237 pages. 1998.

Vol. 1691: R. Bezrukavnikov, M. Finkelberg, V. Schechtman, Factorizable Sheaves and Quantum Groups. X, 282 pages. 1998.

Vol. 1692: T. M. W. Eyre, Quantum Stochastic Calculus and Representations of Lie Superalgebras. IX, 138 pages. 1998.

Vol. 1694: A. Braides, Approximation of Free-Discontinuity Problems. XI, 149 pages. 1998.

Vol. 1695: D. J. Hartfiel, Markov Set-Chains. VIII, 131 pages. 1998.

Vol. 1696: E. Bouscaren (Ed.): Model Theory and Algebraic Geometry. XV, 211 pages. 1998.

Vol. 1697: B. Cockburn, C. Johnson, C.-W. Shu, E. Tadmor, Advanced Numerical Approximation of Nonlinear Hyperbolic Equations. Cetraro, Italy, 1997. Editor: A. Quarteroni. VII, 390 pages. 1998.

Vol. 1698: M. Bhattacharjee, D. Macpherson, R. G. Möller, P. Neumann, Notes on Infinite Permutation Groups. XI, 202 pages. 1998.

Vol. 1699: A. Inoue,Tomita-Takesaki Theory in Algebras of Unbounded Operators. VIII. 241 pages. 1998.

Vol. 1700: W. A. Woyczyński, Burgers-KPZ Turbulence,XI, 318 pages. 1998.

Vol. 1701: Ti-Jun Xiao, J. Liang, The Cauchy Problem of Higher Order Abstract Differential Equations, XII, 302 pages. 1998.

Vol. 1702: J. Ma, J. Yong, Forward-Backward Stochastic Differential Equations and Their Applications. XIII, 270 pages. 1999.

Vol. 1703: R. M. Dudley, R. Norvaiša, Differentiability of Six Operators on Nonsmooth Functions and p-Variation. VIII, 272 pages. 1999.

Vol. 1704: H. Tamanoi, Elliptic Genera and Vertex Operator Super-Algebras. VI. 390 pages. 1999.

Vol. 1705: I. Nikolaev, E. Zhuzhoma, Flows in 2-dimensional Manifolds. XIX, 294 pages. 1999.

Vol. 1706: S. Yu. Pilyugin, Shadowing in Dynamical Systems. XVII, 271 pages. 1999.

Vol. 1707: R. Pytlak, Numerical Methods for Optimal Control Problems with State Constraints. XV, 215 pages. 1999.

Vol. 1708: K. Zuo, Representations of Fundamental Groups of Algebraic Varieties. VII. 139 pages. 1999.

Vol. 1709: J. Azéma, M. Émery, M. Ledoux, M. Yor (Eds), Séminaire de Probabilités XXXIII. VIII. 418 pages. 1999.

Vol. 1710: M. Koecher, The Minnesota Notes on Jordan Algebras and Their Applications. IX, 173 pages. 1999.

Vol. 1711: W. Ricker, Operator Algebras Generated by Commuting Projections: A Vector Measure Approach. XVII, 159 pages. 1999.

Vol. 1712: N. Schwartz, J. J. Madden. Semi-algebraic Function Rings and Reflectors of Partially Ordered Rings. XI, 279 pages. 1999.

Vol. 1713: F. Bethuel, G. Huisken, S. Müller, K. Steffen, Calculus of Variations and Geometric Evolution Problems. Cetraro, 1996. Editors: S. Hildebrandt, M. Struwe. VII, 293 pages. 1999.

Vol. 1714: O. Diekmann, R. Durrett, K. P. Hadeler, P. K. Maini, H. L. Smith, Mathematics Inspired by Biology. Martina Franca, 1997. Editors: V. Capasso, O. Diekmann. VII, 268 pages. 1999.

Vol. 1715: N. V. Krylov, M. Röckner, J. Zabczyk, Stochastic PDE's and Kolmogorov Equations in Infinite Dimensions. Cetraro, 1998. Editor: G. Da Prato. VIII, 239 pages. 1999.

Vol. 1716: J. Coates, R. Greenberg, K. A. Ribet, K. Rubin, Arithmetic Theory of Elliptic Curves. Cetraro, 1997. Editor: C. Viola. VIII, 260 pages. 1999.

Vol. 1717: J. Bertoin, F. Martinelli, Y. Peres, Lectures on Probability Theory and Statistics. Saint-Flour, 1997. Editor: P. Bernard. IX, 291 pages. 1999.

Vol. 1718: A. Eberle, Uniqueness and Non-Uniqueness of Semigroups Generated by Singular Diffusion Operators. VIII, 262 pages. 1999.

Vol. 1719: K. R. Meyer, Periodic Solutions of the N-Body Problem. IX, 144 pages. 1999.

Vol. 1720: D. Elworthy, Y. Le Jan, X-M. Li, On the Geometry of Diffusion Operators and Stochastic Flows. IV, 118 pages. 1999.

Vol. 1721: A. Iarrobino, V. Kanev, Power Sums, Gorenstein Algebras, and Determinantal Loci. XXVII, 345 pages. 1999.

Vol. 1722: R. McCutcheon, Elemental Methods in Ergodic Ramsey Theory. VI, 160 pages. 1999.

Vol. 1723: J. P. Croisille, C. Lebeau, Diffraction by an Immersed Elastic Wedge. VI, 134 pages. 1999.

Vol. 1724: V. N. Kolokoltsov, Semiclassical Analysis for Diffusions and Stochastic Processes. VIII, 347 pages. 2000.

Vol. 1725: D. A. Wolf-Gladrow, Lattice-Gas Cellular Automata and Lattice Boltzmann Models. IX, 308 pages. 2000.

Vol. 1726: V. Marić, Regular Variation and Differential Equations. X, 127 pages. 2000.

Vol. 1727: P. Kravanja, M. Van Barel, Computing the Zeros of Analytic Functions. VII, 111 pages. 2000.

Vol. 1728: K. Gatermann, Computer Algebra Methods for Equivariant Dynamical Systems. XV, 153 pages. 2000.

Vol. 1729: J. Azéma, M. Émery, M. Ledoux, M. Yor, Séminaire de Probabilités XXXIV. VI, 431 pages. 2000.

Vol. 1730: S. Graf, H. Luschgy, Foundations of Quantization for Probability Distributions. X, 230 pages. 2000.

Vol. 1731: T. Hsu, Quilts: Central Extensions, Braid Actions, and Finite Groups,. XII, 185 pages. 2000.

Vol. 1732: K. Keller, Invariant Factors, Julia Equivalences and the (Abstract) Mandelbrot Set. X, 206 pages. 2000.

Vol. 1733: K. Ritter, Average-Case Analysis of Numerical Problems. IX, 254 pages. 2000.

Vol. 1734: M. Espedal, A. Fasano, A. Mikelić, Filtration in Porous Media and Industrial Applications. Cetraro 1998. Editor: A. Fasano. 2000.

Vol. 1735: D. Yafaev, Scattering Theory: Some Old and New Problems. XVI, 169 pages. 2000.

Vol. 1736: B. O. Turesson, Nonlinear Potential Theory and Weighted Sobolev Spaces. XIV, 173 pages. 2000.

Vol. 1737: S. Wakabayashi, Classical Microlocal Analysis in the Space of Hyperfunctions. VIII, 367 pages. 2000.

Vol. 1738: M. Émery, A. Nemirovski, D. Voiculescu, Lectures on Probability Theory and Statistics. XI, 356 pages. 2000.

Vol. 1739: R. Burkard, P. Deuflhard, A. Jameson, J.-L. Lions, G. Strang, Computational Mathematics Driven by Industrial Problems. Martina Franca, 1999. Editors: V. Capasso, H. Engl, J. Periaux. VII, 418 pages. 2000.

Vol. 1740: B. Kawohl, O. Pironneau, L. Tartar, J.-P. Zolesio, Optimal Shape Design. Tróia, Portugal 1999. Editors: A. Cellina, A. Ornelas. IX, 388 pages. 2000.

Vol. 1741: E. Lombardi, Oscillatory Integrals and Phenomena Beyond all Algebraic Orders. XV, 413 pages. 2000.

Vol. 1742: A. Unterberger, Quantization and Non-holomorphic Modular Forms. VIII, 253 pages. 2000.

Vol. 1743: L. Habermann, Riemannian Metrics of Constant Mass and Moduli Spaces of Conformal Structures. XII, 116 pages. 2000.

Vol. 1744: M. Kunze, Non-Smooth Dynamical Systems. X, 228 pages. 2000.

Vol. 1745: V. D. Milman, G. Schechtman, Geometric Aspects of Functional Analysis. VIII, 289 pages. 2000.

Vol. 1746: A. Degtyarev, I. Itenberg, V. Kharlamov, Real Enriques Surfaces. XVI, 259 pages. 2000.

Vol. 1747: L. W. Christensen, Gorenstein Dimensions. VIII, 204 pages. 2000.

Vol. 1748: M. Růžička, Electrorheological Fluids: Modeling and Mathematical Theory. XV, 176 pages. 2001.

Vol. 1749: M. Fuchs, G. Seregin, Variational Methods for Problems from Plasticity Theory and for Generalized Newtonian Fluids. VI, 269 pages. 2001.

Vol. 1750: B. Conrad, Grothendieck Duality and Base Change. X, 296 pages. 2001.

Vol. 1751: N. J. Cutland, Loeb Measures in Practice: Recent Advances. XI, 111 pages. 2001.

Vol. 1752: Y. V. Nesterenko, P. Philippon, Introduction to Algebraic Independence Theory. XIII, 256 pages. 2001.

Vol. 1753: A. I. Bobenko, U. Eitner, Painlevé Equations in the Differential Geometry of Surfaces. VI, 120 pages. 2001.

Vol. 1754: W. Bertram, The Geometry of Jordan and Lie Structures. XVI, 269 pages. 2001.

Vol. 1755: J. Azéma, M. Émery, M. Ledoux, M. Yor, Séminaire de Probabilités XXXV. VI, 427 pages. 2001.

Vol. 1756: P. E. Zhidkov, Korteweg de Vries and Nonlinear Schrödinger Equations: Qualitative Theory. VII, 147 pages. 2001.

Vol. 1757: R. R. Phelps, Lectures on Choquet's Theorem. VII, 124 pages. 2001.

Vol. 1758: N. Monod, Continous Bounded Cohomology of Locally Compact Groups. X, 214 pages. 2001.

Vol. 1759: Y. Abe, K. Kopfermann, Toroidal Groups. VIII, 133 pages. 2001.

Vol. 1760: D. Filipović, Consistency Problems for Heath-Jarrow-Morton Interest Rate Models. VIII, 134 pages. 2001.

Recent Reprints and New Editions

Vol. 1200: V. D. Milman, G. Schechtmann, Asymptotic Theory of Finite Dimensional Normed Spaces - Corrected Second Printing 2001. X, 156 pages. 1986.

Vol. 1618: G. Pisier, Similarity Problems and Completely Bounded Maps - Second, Expanded Edition VII, 198 pages. 2001.

Vol. 1629: J. D. Moore, Lectures on Seiberg-Witten Invariants - Second Edition. VIII, 121 pages. 2001.

Vol. 1702: J. Ma, J. Yong, Forward-Backward Stochastic Differential Equations and Their Applications - Corrected Second Printing 2000. XIII, 270 pages. 1999.

GPSR Compliance

The European Union's (EU) General Product Safety Regulation (GPSR) is a set of rules that requires consumer products to be safe and our obligations to ensure this.

If you have any concerns about our products, you can contact us on ProductSafety@springernature.com

In case Publisher is established outside the EU, the EU authorized representative is:

Springer Nature Customer Service Center GmbH
Europaplatz 3
69115 Heidelberg, Germany

Batch number: 09301484

Printed by Printforce, the Netherlands